NOW I CAN TELL

NOW I CAN TELL

The Story of a Christian Bishop Under
Communist Persecution

QUENTIN K. Y. HUANG

MOREHOUSE-GORHAM CO.
New York
1954

Printed in U. S. A.
By The Haddon Craftsmen, Inc., Scranton, Pa.

Dedicated To

GOD

Who Chastens Those Whom He Loves
and to
Those Who Obey God Rather Than Man

Publisher's Foreword

THIS is the first book written by a Chinese clergyman on applied Communism and the drastic Communist land reforms behind the Bamboo Curtain in Red China, based on his personal experiences and sources not accessible to others. Bishop Huang knows what he is talking about for he was a prisoner of the Chinese Communists and gained his knowledge and insight into their aims and methods the hard way—through imprisonment, starvation, torture, and attempted brain washing.

The author of this book, the Rt. Rev. Quentin K. Y. Huang, is a Bishop of the Holy Catholic Church in China— the Chinese branch of the Anglican communion. The diocese of Yunkwei, of which he was the head, is in Yunnan and Kweichow, Southwest China, and it was there that he was arrested and imprisoned by the Chinese Communists. Since his escape he has been a refugee in the United States and associate rector of the Church of St. Stephen and the Incarnation, Washington, D. C.

Bishop Huang is a graduate of St. Paul's School, Anking, and St. John's University, Shanghai, both missionary institutions founded by the American Episcopal Church. He received the degree of Master of Arts from the University of Pennsylvania and those of Bachelor and Master of Sacred Theology and Doctor of Divinity from Philadelphia Divinity School. Returning to China after ordination, he was a professor at a number of colleges and universities in his country and also a pastor of the Chinese Church.

Elected Bishop in 1946, he returned to this country and was consecrated to the episcopate in California. He then returned to his own country, where he organized his diocese and was endeavoring to put it on a self-sustaining basis when the Chinese Communists took over the province in which he had his jurisdiction.

Although he attempted to remain aloof from political activities, Bishop Huang soon fell into disfavor with the Communist authorities and was arrested and imprisoned on trumped-up charges. Part I of this book tells his personal experience with the Chinese Communists, including his imprisonment, his offer of high position if he would submit to the Communists, his refusal, and his ultimate escape. It gives a unique insight into the subtle methods and real objectives of Communist policies and procedures, and is a vital witness to God's power and grace in his loyalty under extreme pressure.

In Part II the real objectives of their Land Reform Program are told in detail. This section contains the careful study of the Communist Land Reform Program which Bishop Huang was able to make, and which is here published for the first time. It shows conclusively the way in which this apparently democratic movement, which so long led Western observers to think that the Chinese Communists were liberal agrarian reformers, is actually a powerful instrument to deceive and subjugate the masses, and to consolidate Communist power for world revolution.

THE PUBLISHERS

Preface

To MY FRIENDS it seems strange that I, as a strong believer in, and advocate of, separation between Church and State, should speak and write on APPLIED COMMUNISM. Or they may think that the reason I so speak and write is due to some desire for revenge for the injustice and imprisonment I suffered at the hands of the Communists.

No, absolutely no! Instead, I am very grateful to God for the lessons I learned through suffering. I still believe that God is Love, that He loves His rebellious children just as much as His obedient ones, and that He will in due time remove this menace to mankind, if God-fearing people coöperate with Him. Neither for any interest in politics nor for any grudge against any person am I against the sin and evil of Communism, but because I am convinced that this present world-conflict is spiritual. It is an ideological conflict between Materialism and Christianity, between Communism and Democracy. "Marxism," as Lenin stated, "is Materialism militant," or, as the leading Chinese Communists have repeatedly expressed in their writings, "The Chinese Communist revolution is a part and parcel of the world revolution."

Until the whole world is revolutionized, or Communism itself is totally destroyed, it will hoist its flag of "Classless Utopia" and be followed by its materialistic fanatics, who believe in no god but the Party-State—the major premise of their dialectical and diabolical system of philosophy. To that end, everything is a tool and every person a slave to be utilized, not sanctified. The denial of God as Father of all logically winds up in the denial of individual dignity, personal rights, and the brotherhood of man. The repudiation of God's sovereignty leads rationally to the repudiation of morality and moral values, and justifies the belief that "the means is justified by the end."

Because of its materialistic philosophy, Communism leaves no room for any idealistic systems of thought, condemns religion as the "opiate of the people," allows no compromise, and constitutes the greatest threat not only to our democratic way of life but also to all cultures of mankind founded on the idealistic or spiritual basis. The Communists believe that they alone have Truth, and nobody else; so, in both theory and practice, all those who do not go along with them are their enemies. "Inability to distinguish between enemies and us [the Communists] is both a crime and sin," for which countless numbers of our fellow-countrymen and Christian brethren have been condemned.

On the other hand, we Christians believe that men are children of God, not only as physical beings but also as spiritual entities; not merely as animals but also having souls, capable of reasoning, capable of listening to the inner voice (conscience), capable of choice, capable of judgment, and capable of developing and perfecting themselves after the Image of God. This concept of man leads to the reality of God, the dignity of the individual, the brotherhood of man, and the Kingdom of God. To say nothing about the differences in the methods of Communism and Christianity, these Christian beliefs and those of materialistic Communism are absolutely as opposite as the two poles. To bring them together, reconciled or compromised, I have seen no light. We cannot interpret Marxism by Christianity, as the Communists would not tolerate it; nor can we interpret Christianity by Marxism, as a few of our Christian brethren in China have attempted. There will be neither "Christmarxism" nor "Marxianity." In this ideological conflict between Christianity and Communism, there is no compromise possible, and our Lord has warned us by saying, "He who is not with me is against me."

Lately, one problem in the free democratic countries has puzzled me greatly. Many patriots who are strong defenders of democracy, willing to do everything possible to build a strong democracy, pay no attention whatsoever to religion or their faith, the real foundation of democracy! To me, they are merely trying to build "air castles." We cannot love our neighbors as ourselves, if not through Christ. We cannot regard all men as equal, if we do not see them as children of God. We cannot consider them as brothers, if we reject the Fatherhood of God. We cannot have individual dignity and rights, if we are mere animals and not souls created in the Image of God. We cannot build a castle without a foundation, nor can we plant a fruit tree without soil. So we cannot build a strong democracy by detaching it from its religious foundation, its motivating ideology and its soul. Arnold Toynbee concluded his conviction on this subject by asking, "How can the western nations successfully combat Communism unless they establish an active, working Christianity?"

If we believe conscientiously in the reality of God, the dignity of the individual, and the brotherhood of man—the religious foundation of democracy—we cannot remain neutral or above politics, as many of us would like to be, because this conflict is not only political, economic, and national, but also ideological, spiritual, and international. We are in the conflict, whether we like it or not. We may be discontented or even disgusted with the realities and injustices of this phenomenal world, but we must recognize that they are the results of our human weakness and failure. This, I believe, if not the main cause, is at least one of the main causes for Communism coming into existence and becoming such a predominant force in the world today.

To combat Communism, it is therefore the bounden duty of every God-fearing person, and every human being living and enjoying personal rights and freedom in free countries, to regenerate himself and put his faith into action so as to build a strong foundation for democracy, instead of playing or experimenting with something new and novel, clever in propaganda and sweet in promise, and dressed in sheep's clothing, not recognizing inwardly the ravening wolf itself. It is to expose this inward ravening wolf, Applied Communism, behind the Bamboo Curtain (which is totally different from theoretical Communism or New Democracy), that I have undertaken the task of writing this book so as to fulfill, I am convinced, one of my duties in this great spiritual conflict today.

By the grace of God I was arrested, imprisoned, indoctrinated, and forced to go through both mental and physical sufferings and the Communist procedure of remaking human beings. It was through His love and mercy that I, though one of His unworthy servants, was able by means of medical and other services to cultivate and develop intimate friendship with the Communist agents imprisoned to spy on us, as well as with our fellow-prisoners, some of whom were exceptionally rich in experience with Chinese Communism. Through our intimate friendship, I gained knowledge and insight of the subtle procedures and real objectives of Chinese Communism which no foreign observer or prisoner of Red China would be able to gain from what he is allowed to see, hear, or read, or a native without exceptional opportunities would be able to learn, as the Chinese Communists plan things behind curtains and carry out projects in the dark.

One of the Chinese proverbs says, "To know thyself and thy enemy is the secret to victory in battles." In combatting the evils of Communism, we must know thoroughly both ourselves and our adversaries. During the past couple of years we have read a few good articles on "Communism Behind the Bamboo Curtain" and a few books on the Chinese Communist indoctrination, but they have given us either only a very small portion of the picture or what they have tried to paint is too superficial. I have waited two years for some keen observer to tell the whole story, but have found none satisfactory. I myself could not have done it earlier for the simple and single reason of protecting the security of our fellow-Christians in China. But NOW I CAN TELL, because the issues are clear. During the period between October, 1949, and the end of 1952, the Communist regime in China killed 186,069 Christians, and still has 390,420 imprisoned. Many of our own Christian friends have been arrested, and imprisoned or killed. As far as I myself am concerned, it is definitely known, even to the Communist authorities in Peiping, that my direct connections with the Church and its leaders there have been cut since my escape and that we have been on our own, independent and self-responsible. Urged by my conscience and encouraged by many friends from time to time, I feel it is both my obligation and privilege that I should tell my experiences with the Communists behind the "Bamboo Curtain."

Based entirely upon, and limited to, my personal experiences and knowledge of Chinese Communism, this book is divided into two parts. The first part is really my own story, or rather, a series of stories of experiences with the Communists, arranged chronologically. Read separately, each chapter is a complete story in itself. While taken as a whole (with a little repetition, naturally, here and there), it is the story of my imprisonment and escape, a real and conscientious witness to God's power and grace. To safeguard some of my friends and their beloved ones still behind the curtain, I have purposely omitted some names, or at other times used only fictitious ones, as indicated, and have kept some places indefinite. Again, intentionally, I have left out everything possible pertaining to the Church behind the Bamboo Curtain, but some day, if God permits and inspires, I may take this up as my second adventure.

The second part concerns Communist land reforms in China, including their subtle procedures and real objectives. This is really the key to the understanding of Applied Communism in China, how it works, enslaves the people, consolidates its power, etc. This should be read with care and imagination so as to realize how treacherous and inhuman the Communists are to the masses of people in China today. It has happened in China and it may happen elsewhere if we believe in their sweet words and think that Communism is one of the liberal movements today.

I hope and pray that my readers, after having learned the true nature of this inward ravening wolf, may be aware, on the one hand, of the impossibility of making compromises with our common human adversary, and, on the other, may pray more, give more, work more, and live up more to the faith of their choice—the fountainhead of democracy and the spiritual weapon with which we may arm ourselves in this great ideological warfare of ours.

The quotations in Part II, except those indicated, are from a confidential source, to whom I hereby express my gratitude.

I am deeply indebted to Mrs. Henrietta Geyer, who has done more than I can well express to help prepare my book and make it ready for early publication.

Last, I wish to take this occasion to express my deep appreciation to all my friends throughout the States, both old and new, who have taken a keen interest in my experiences and my book, particularly to Miss Florence L. Newbold, Mr. Richard Whittington, Mr. Robert B. Reed, Mr. F. Stark Newberry, Dr. Wayne Y. H. Ho, Dr. Clifford P. Morehouse, the Rev. B. DeFrees Brien, the Rev. C. Capers Satterlee, the Rev. Arthur K. Fenton, the Rev. Matthew H. Imrie, the Very Rev. J. Milton Richardson, the Rev. Sidney M. Hopson, the Rev. Stuart F. Gast, and the Rt. Rev. William S. Thomas, Jr. They have helped me by making it possible for me to bear my witness to our Lord with my experiences. Above all, I am most grateful to my wife for her love, devotion, encouragement, inspiration, and particularly for her willingness to share all my headaches and sufferings both during the days of distress and in our united struggle for Truth.

QUENTIN K. Y. HUANG

Washington, D. C.

Contents

Publisher's Foreword vii

Preface ix

I. MY EXPERIENCES WITH COMMUNISTS BEHIND THE
BAMBOO CURTAIN:

  1. A Black Cat Turned Out to Be a Blessing of God   3
  2. How I Became the First Bishop to Be
     Imprisoned 7
  3. Locked Up in the Wooden Cage 16
  4. Life in the Wooden Cage 22
  5. Parade of Prisoners 32
  6. God's Providence and My Religious Work . . 37
  7. Picked as an Example of "The Enemy" to Be
     Reported 46
  8. Communist Blackmail 53
  9. Communist Care vs. God's Love and Miracle   59
 10. Organizing "Study Groups" for Indoctrination   69
 11. Communist Grace and the Prisoners' Agony   74
 12. Communist Requirements Before Trials . . . 80
 13. Communist Judges and Trials 86
 14. My Accusers and Second Trial 94
 15. Intensive Indoctrination 102
 16. Communism and Children 114
 17. Class Classification—Our Headache 118
 18. My Trial at Night: How Truth Nearly Killed,
     but Saved Me 125

19. Subtlety of Communism 134
20. A Student for the Ministry Who Worked for the
Devil 140
21. My Release and Bewilderment 151
22. House Arrest and Welcome Party 159
23. God's Will Finally Revealed 165
24. My Escape 171

II. COMMUNIST LAND REFORMS IN CHINA:
Subtle Procedure and Real Objectives . . . 185
The Communist Version of Eminent Domain 220

Part I

MY EXPERIENCES WITH COMMUNISTS
BEHIND THE BAMBOO CURTAIN

1

A Black Cat Turned Out to Be a Blessing of God

OUR DIOCESE was named Yunkwei, as it took in all the territory of two provinces, Yunnan and Kweichow, in Southwest China. Yunnan is in the west while Kweichow is in the east. Work was almost equally divided between these two provinces, half of the colleagues being stationed in Kweichow under the charge of the newly appointed Archdeacon.

The Communist forces were sweeping from east to west in the fall of 1949; soon Kweichow was under the claws of the Red and fenced off by his Iron Curtain. We in Kunming, Yunnan, the see city of our diocese, were entirely cut off from our fellow workers in Kweichow. Rumors spread like wildfire, thousands were imprisoned, and hundreds brutally massacred in the cities and towns of Kweichow. Day in and day out, we all went out in the daytime, trying to make some contacts with our workers in Kweichow by means of telegrams or radiograms, and at night we gathered together in the Diocesan House and prayed for them. But it was all fruitless and hopeless!

Almost two months went by and we had not had a word from any one of our workers. Whether they were living or dead was the biggest question. If still alive, what could we do for them? Anxiety piled up daily! At last, on the 28th of

November, 1949, I called an emergency meeting of the members of the Diocesan Standing Committee. The main subject on the agenda was the work and the workers in Kweichow. With a heavy heart, every member of the committee came to the meeting but, with the help of God, a wise resolution was unanimously passed that a Yunkwei office be established in Hongkong, a British colony, from where our workers in Kweichow might be contacted. As discussion went on, more serious problems were raised one by one. Rumors and stories had been circulated that the first action of the Communist regime would be to decentralize the Church organizations and localize the ecclesiastical authorities; that English would be forbidden to be used, and that funds or financial assistance from abroad would be cut off, etc.

To cope with this delicate and complicated situation, whether true or false, and also with a view to the possibility of the turning over of the Yunnan provincial government to the Communists, it was decided that a competent representative, and Chinese, should be chosen, one who had an intimate knowledge of the characteristics and problems of each individual worker and was capable of making connections in Hongkong so as to send assistance to our workers, and who also had the knowledge of our Yunkwei patrons abroad. (I had decided that in this time of emergency I should remain in order to give peace and consolation to our colleagues. I had written and asked the House of Bishops in China for permission to be absent from its called emergency meeting to be held in Shanghai the first part of December.)

Going over the whole list of workers for a possible candidate for this important and complicated post in Hongkong, we found none except myself and, possibly, my wife, who was not a paid worker and, therefore, the Diocesan Standing

Committee had no right to send her. But after much discussion, one member moved that my wife, whether the committee had the right or not, be requested to go down to Hongkong and establish the Yunkwei office there, for the sake of the work and the workers. With much hesitation, I let the proposal be seconded and passed.

After the meeting, the request was presented to my wife. She unhesitatingly turned it down flat, saying, "In the time of emergency, as was true many times in the Second War, my husband and I have always been separated. This time we have decided to remain together for better or for worse, and we will stay together. Please get somebody else!" I was then on the horns of a dilemma, much confused, and did not know what to do. For the sake of God's work and the workers, I felt it was my duty to ask her to go, as she was the one person in the whole diocese qualified to render that complicated and important service. On the other hand, in a time that was so uncertain and critical—no one could foretell what would be next—I personally agreed with her on staying together. I was indeed at the crossroads, particularly when she was so determined that no one could even discuss the subject before or with her. In my bewilderment, I asked all concerned to pray more about it.

This unhappy situation worried her a great deal, so much that at night her old illness, asthma, came back to her. She had to stay in bed resting, but was really restless. While annoyed at me and the members of the standing committee, she was a devout woman and prayed for her recovery and God's guidance in this muddle. Three days later, her asthma was gone and she was on her feet again. Looking rather happy in the bright morning sunshine, she commented, "I guess I have to listen to God rather than to my own wish, but this is the last service and there is to be absolutely no more separation in any emergency!"

"This illustrates," I said, "that with man it is impossible to change your mind, but not with God."

So she began to pack while I went out to make reservations. At 9:00 a.m. on December 3rd I was at the ticket office and found that all tickets for the flights on December 4th had been sold. Just when I was about to decide to get tickets for her and Joy Ann, our daughter in Kunming, for the next flight on December 6th, two men came in and expressed their deep regret for the necessity of postponing their flight to Hongkong. The booking authority consented and happily turned over the tickets to me. So the next day my wife and daughter flew down to Hongkong. This proved to be the last commercial plane. After that, the military authorities in Kunming seized every plane, both commercial and military, for their own use.

Five days later, on December 9th, Lu Han, Governor of Yunnan, surrendered himself to the Communist regime. It was only by the providence of God that my wife left, and left in time. Her departure, at the time a "black cat" to us, later turned out to be a great blessing of God. ("Black cat" in China means misfortune and poverty, as one of the Chinese sayings states, "Cats bring poverty, while dogs bring prosperity.") By the grace of God, my escape later was made much easier!

O God, how mysteriously Thou didst prepare and work for the care and protection of Thy unworthy servants!

2

How I Became the First Bishop to Be Imprisoned

THE HUEITIEN HOSPITAL of our Church, with 130 beds, 20 doctors, and 40 nurses, was considered the best and largest in Southwest China. It was originally started by missionaries of the Church Missionary Society, full of the Christian spirit of love and service. At one time, it was an effective means through which many Chinese were brought to Christ, but later on, human weaknesses—jealousy, selfishness, and greed—crept in. Slowly it became no more a means of Christian grace and evangelism, but a hotbed of worldliness and a struggle for power. The hospital authorities treated it as a secular enterprise, attempting constantly to make it independent, while the Church authorities, sharing an equal if not greater responsibility, step by step lost their grip on it. By the time I took over the episcopate in 1946, I was confronted with more difficulties from the so-called Christians than from the "heathen brethren," as Mr. and Mrs. King, two leading Christians in Kunming, often commented. However, they encouraged us by saying, "You are stepping into a pair of thorny shoes, but we'll help!"

Because of the secular and decaying status of this institution, both the card members and the associates of the Communist Party easily infiltrated the hospital as servants, attendants, students, nurses, and doctors. They were constantly making trouble, demanding higher salaries and wages, better food and living quarters. There were quarrels and fights between cliques—headaches after headaches! By patience in dealing with the workers, special emphasis on the importance of service to the sick, alertness in watching the actions of those reported as being "red" or "pink," and attempting to nip their plots in the bud; by giving every branch of the hospital staff proportional representation and voice in the allotment of apartments and rooms, in the management of food and in the scale of salaries and wages; by giving every worker the right and privilege of looking into the hospital finances and accounts; and by enforcing impartial treatment to all without exception, we were able to keep both the school of nursing and the hospital going before the Communists took over Kunming. Although the Communist infiltrators tried several times to wreck both institutions, they could not swing the majority in order to fulfill their desire and hope. At last they changed their tactics! Instead of wrecking the hospital and the school of nursing to pave the way for the new regime to take over, they began to attack the superintendent and me soon after the city had fallen into the Communists' hands.

One week after the surrender of Lu Han, Governor of Yunnan, to the Communists (on December 9, 1949), there was a quarrel between the superintendent and Dr. Fu, head of the Department of Medicine. In addition to scolding each other, the superintendent warned the physician that he would be discharged should he interfere with the administration again. In reply, Dr. Fu, who was an associate member of the Communist Party, carelessly and angrily commented, "You discharge me? You just wait and see whether I shall be discharged by you, or you by me!" Dr. Fu, although he had no connection with our school of nursing, participated in every Communist parade and gathering by leading the student nurses. He was soon made head of the Medical Association in Kunming by the Communist authorities.

Twice before and once after the change of government, I received letters through the mail from the Communist organization, denouncing me as a "running dog" of American imperialism and threatening me with imprisonment and death. However, with the proclamation of freedom of religion by the Communist government in Peiping, and my own conscience being clear, I did not pay any attention to these threats.

On the morning of December 26, 1949, after the chapel service in the hospital, Dr. Henry P. Brown, of Philadelphia, who had volunteered himself to help with our medical work in Yunkwei, wanted me to go to his house with him as he had something important to ask me. Sitting in the parlor, he looked rather serious and started to question me: "Quentin, are you in business?"

Surprised by his question, I said, "Why? I should say, *NO*. I have never been in business in my life. Of course, I have bought lots of things, but have never sold anything. For instance, we bought a lot of cotton yarns from the Kunming Mill, through Mr. King, in order to keep the value of currency from depreciating, but we divide the yarns among the Church workers as their salaries. It is up to each individual worker to sell or to keep the yarn, as he chooses. That is the decision of the Standing Committee, in order to safeguard the value of the workers' salaries."

"Have you bought lots of medicine?" he questioned again;

"Or have you sold any medicine from your diocesan drug factory?"

"Yes," I answered; "Mrs. Phinney (the missionary sent over by the Girls' Friendly Society in the United States to help us in Yunkwei) is trying to buy some in Hongkong, partly for the hospital and partly for the diocesan drug factory. Our drug factory is still in the experimental stage; whether we shall be able to make really good medicine or not is still a question. What we are doing now is to fill the prescriptions of the hospital. Of course, we have sent a lot of medicine to the hospital, but we have not received a cent for it. Furthermore, we have not decided whether we should try to get a license from the government for the factory. [Note: The diocesan drug factory was one of our self-supporting projects.] Unless we get such a license, we cannot sell what we make to anybody, so it still remains the dispensary of the hospital."

"What will you do with those five truckloads of medicine brought over from Kweiyang by Dr. Chia, Dr. Chow, and yourself, and at present stored in the hospital?" inquired Dr. Brown. "I have been informed that it is known in medical circles in Kunming that you have coöperated in this illegal dealing with Dr. Chia and Dr. Chow for the purpose of buying them cheaply and selling them later at a great profit. Isn't it true?"

I was shocked and dumfounded to hear about the five truckloads of medicine and said, "For heaven's sake, I don't know a thing about the medicine. During the emergency in Kweichow Province, I was in Kweiyang planning some emergency measures for our Church workers there—you know it! On my way back, Mrs. Phinney and I traveled together; you can ask her whether there was any medicine with us or not. We traveled by bus and were certainly limited as to luggage, so Mrs. Phinney had to leave lots of her things in Kweiyang. If we had had five trucks loaded with medicine, we could and would have brought her belongings over to Kunming. It is a plot, I know! I will have to find out about the whole business and straighten it out as soon as possible; otherwise, I would be involved in politics and in buying government medicine illegally, which is quite serious, especially at a time like this! Who told you such a big lie, may I ask?"

"Never mind about the person. I'll not tell you," answered Dr. Brown.

I left immediately and went over to see Dr. Chow, the head of our new department, and our superintendent. From them I learned the whole story. There were five trucks loaded with medicine belonging to the Kweichow provincial government. Before the evacuation of the Kweichow government, Dr. Chia, head of the Kweichow Provincial Health Bureau, was ordered by Governor Ku of Kweichow to have the medicine transported to Kunming. "The man in charge of the transportation is still in Kunming," said Dr. Chow, "and I will get him to come and testify."

"For the sake of courtesy," explained the superintendent, "I consented to have the medicine stored in our hospital temporarily. It is piled up in a separate corner of the warehouse. Yesterday the Public Safety Bureau of the new regime sent two men over to see me and asked me about the medicine, so I told them about it. They are supposed to come again tomorrow and take it away."

After a short consultation, it was decided that an emergency meeting of all the members of the hospital staff would be held on Thursday, December 28, at 7:30 p.m., at the mess hall. A large notice of the meeting was immediately posted on the bulletin board, with no mention of its purpose. When the time came, the hall was full of people, and the man in charge of the transportation of those five trucks of medicine was brought in. After his testimony, both Dr. Chow and the superintendent told the meeting the whole story—how it got to the hospital warehouse, etc., and that it had nothing to do with me. Those Communist infiltrators, sitting in the meeting and listening to the testimony and reports, felt very much ashamed and embarrassed, for their plot had come to naught. After the meeting, one fellow came to me, whispering, "We were surprised when we heard the rumors spreading around the hospital, but I am glad you have cleared yourself tonight!"

I thanked God for giving me the chance of pulling myself out of this trap in time; otherwise I would have been involved in an illegal political and secular affair and would have been embarrassed and condemned, without an opportunity of clearing myself before my friends and fellow workers. Lies often repeated and circulated, half truths and truths twisted, were some of the weapons of the Communists by which others might be brought to submission or liquidation.

The failure of this plot for spoiling my reputation and bringing me to submission speeded up my arrest. The meeting was reported by the Communist agent and only four hours later, at 2:00 a.m. on December 29, a group of Communists from the Bureau of Public Safety went to my residence in the Diocesan Center and found that I was not at home.

During those days of arrests (the reign of terror), the night curfew was still in effect between 8:00 p.m. and 6:00 a.m. After the meeting at the hospital, it was too late for me to go back to my house, so the superintendent put me up for the night in his study. He had no extra bed but had a mattress, which I used to sleep on the floor. It was a night of annoyance because of the rats, some dashing from corner to corner, others dancing around, and a few running over my face. I had to get up once every half hour to switch on the light in order to check them.

Early next morning I trudged back to the Diocesan Center, without washing, and heard that a group of Communists from the Bureau of Public Safety had come at 2:00 a.m. to inquire about the "Head." With my conscience clear, I did not pay much attention to the report and went to the parlor to try to get some rest.

At 10:00 a.m. on the same day, the same group came back again, consisting of two uniformed men from the Public Safety Bureau, two policemen from the Fourth Police Station, Mr. Ma Pao Chang (the sectional leader of 100 families), and a youngster of 16, also in uniform, with a pistol hanging down on his left side. Very politely the head of the group asked, "Where is Quentin Huang?"

Meeting them myself in the front yard of my house, I answered, "It is I. Come in and sit down." They were welcomed into the parlor and entertained with tea. After a while, the leader spoke very courteously again, "We are here just to have a look at where you stay and what you have in your house."

"You are welcome to look anywhere you like," I replied.

"Where is your bedroom?" he asked.

"Upstairs," I answered. "As I am staying alone, I occupy only one room upstairs and all the other rooms have been assigned to our theological students," and led the way.

In my bedroom the four of them searched and searched

everywhere; every box was opened, every cloth unfolded, every corner looked into, and every word of my correspondence carefully read. When they found my portable typewriter (a gift from the Rev. Dr. Arthur Sherman of New York in 1947), curiosity and seriousness were manifested on their faces. Questions were asked, one after another, without waiting for answers. "What is this?" "Is this a radio transmitter?" "Is this a receiving set?" It took me at least half an hour to explain and demonstrate the use of a typewriter.

Soon their fingers were laid on a light meter on the bureau, and my troubles began again. They insisted that it was another radio transmitter. Again, I had to explain and demonstrate. As a result, nothing illegal was found. When we got downstairs they came into the parlor and we talked for a few minutes. Then, very courteously (more so than anybody would expect), the head of the group pulled out a piece of printed paper, filled in with my name and signed by the newly appointed Commissioner of Public Safety, Sung Yi Heng. It was an order of arrest, but he explained to me by saying, "This order is nothing. We have not found anything in your place and we don't expect to find anything. Just take this order and go to the Fourth Police Station, because the Commissioner of Police would like to have a talk with you. You may explain both the typewriter and light meter to him. Everything will be all right."

We walked out together. As soon as we got outside the Diocesan House, Ma Pao Chang returned to his house nearby and the three in uniform left for the mysterious Bureau of Public Safety (the location of which was still unknown to the public), while I was escorted by the two policemen, one on either side, to the Fourth Police Station. I felt hungry and fatigued, but my mind was at peace. We trod along in the bright Yunnan sunshine and, twenty minutes later, we arrived at our destination. I expected to have the case explained and clarified with the Police Commissioner in an hour or so, but instead of seeing the Commissioner myself, I became a captive of the Communists, and the first bishop to be imprisoned by them in China.

3

Locked Up in the Wooden Cage

ARRIVING at the Fourth Police Station outside the West Gate, the first thing I heard was the yelling of the head guard to his subordinates, "Here comes one of the biggest criminals; put on double guard." Unconscious of the fact that the order applied to myself, at first I was also trying to find out who the biggest criminal was. Pretty soon all the eyes in the wooden cage, as well as those in the Rest Room, were focusing upon me and I realized that I was the center of attraction and object of their curiosity. I myself was the biggest criminal! Immediately I asked myself, "*Am* I one of the biggest criminals? What have I done? It is really upside down now. By what standard am I judged? Possibly our Lord was right in foretelling us, 'Because ye are not of the world, but I have chosen you out of the world; therefore, the world hateth you.' So I am the biggest criminal!" This was the beginning of my suspicions regarding the attitude of the new regime toward the people and particularly toward the Church. A few minutes later I pulled myself back to normality by saying to myself, "With my conscience clear before God and man, whom shall I fear?"

Soon I went over to the man in charge, who was dressed in the new, popular Lenin uniform, and courteously asked him, "May I see the Commissioner, as promised, and have my case clarified, sir?" Without looking at me, he, however, answered politely, "Your case has been reported to the Commissioner. You just wait there." Ten minutes later, a guard came in and started taking away all my personal belongings, including glasses, fountain pen, pencil, Ronson lighter (a present from a GI in World War II), watch, pocketbook, note book, bunch of keys, pocket knife, one pair of sleeve buttons, belt, pair of shoe laces, etc. No talk with the Commissioner or anybody else! No argument! No plea! I was just pushed into the wooden cage and locked up!

In the meantime, there was a rich young man, rosy and fat, well-dressed in Western clothes, jumping up and down in the wooden cage, beating his chest and scolding wildly at the injustice of being arrested for no cause or charge. His name was Hsiao Chieh, a native of Hupeh. He had been locked up in the wooden cage seven or eight hours earlier. Hearing the head guard's cry, "Here comes one of the biggest criminals; put on double guard," he was surprised to see me and anxious to find out how great a criminal I was, what I had done, and how I had been arrested. All such questions ran through his mind. He became quiet for a while and was trying hard to find the answers by searching me from head to feet with his glittering eyes. Before he could solve the puzzle, he found that I was ordered and pushed into the same wooden cage. Opening up the door of the cage and looking at the well-dressed young man, Hsiao, the guard commented laughingly, "Here comes another member of your party." This enraged Hsiao. He jumped up again and questioned him, "Which party? Which party? I don't know this man. You—you dogs, you just label us at your will! Damn you!"

Entering the wooden cage, doomed for trouble and suffering, I was really feeling blue and perplexed. All my thoughts were concentrated on myself, and how to solve my own problems and difficulties, and get out—no time for anything or anybody else! Suddenly a beam of light, flashing over my eyes as if somebody had turned a flashlight on me, seemed to say, "Remember, you are a Christian and the head of the Church here. Your business now is to calm and strengthen this weak brother of yours!" Bowing down my head, I prayed mentally, "O Lord, I am troubled and have no peace myself; how can I calm him? I am weak myself; how can I strengthen him? Give me peace and power so that I may be made worthy!"

The wooden cage was very small, only six by eight feet, and 18 persons were packed solid together in it. Hsiao and I were the last two put in; consequently, we two were closest to each other near the cage door. As always among fellow-sufferers, sympathy and intimacy grew quickly between us. Before I entered the cage he had learned from a guard whispering to him that I was a bishop of the Holy Catholic (Episcopal) Church in China. I looked at him several times, trying to get a chance of speaking to him.

Before I had the opportunity he started without any introduction, in a hoarse voice, "Hey, Nan-yu (or suffering friend, the common name by which we prisoners called one another), are you really a bishop?" "Yes," I responded, "I am the bishop of the Holy Catholic Church in Yunnan and Kweichow. For what I am charged, I don't know. I am ashamed to be here but I believe God is justice and love. He will pull us through and declare us innocent! Say, Nan-yu, I saw you jumping up and down, beating yourself and condemning everybody else. What's the use of doing so? You cannot do anybody any harm except yourself. We are locked up! Although we cannot do anything ourselves, remember, there is God who can! Do you believe in God?"

To my happy surprise, I found that he was a Christian, baptized in the London Mission in Hankow, Central China. He had known both Bishop Roots and Father Wood of our Church there in his childhood, as his father often invited them to his house for conferences on relief work.

He then told me confidentially, in a small voice, "My elder brother was working in the Nationalist government in Shanghai and had some connection with the Communist Party, so he is able to remain there working for the New Regime. My second brother is a merchant in Hongkong in the import and export business. I have been in Hongkong helping my second brother but, on account of unstable conditions in the Southwest, I flew up to Kunming on December 1 for the purpose of collecting funds and winding up our branch office. I worked fast and chartered a plane to fly back to Hongkong on the morning of the 9th. Unfortunately, Lu Han, Governor of Yunnan, had revolted against the Central government and turned to the Communists just a few hours before my departure. Consequently, my plane was seized and all my belongings were confiscated.

"Since then I have been staying with one of my friends. In the daytime I have gone around to pull all the strings we have in order to get back my belongings, but at night, I must confess, I have led a wild life with a number of my newly made friends and the Kunming girls. I know them all by now, and there are indeed a few beautiful girls! I have had quite a number of fights over two of them with those rotten and cowardly Kunming friends of mine. Probably, due to jealousy, one of them dropped my name into one of the five secret-reporting boxes.

"At one o'clock on the morning of December 28, when we were having a grand time with two most beautiful girls playing Ma-chiang [a Chinese game, also called Square Wall Battle], three young fellows in Lenin uniform, one of whom was a rival of mine, came to my friend's home and arrested me on an order issued by Sung Yi Hang which mentioned no charges against me. So here I am, suffering and celebrating the Communist Regime. It is unlawful, unjust, and disorderly!"

Suspicious of his telling the truth, I commented in a friendly way, "You must have had lots of personal belongings to be able to charter a plane, and they must be valuable, too." "Oh no, not much, but valuable, though," he answered quickly.

"Friend, forget material things. No matter how valuable, they are nothing. Do you remember the rich fool?" I asked. "That is a good lesson for us today. I am glad you are a Christian and we can talk in our own language. There is no use to worry or get mad. Although we are cut off from our friends and relatives, we are not alone. We are helpless in ourselves but not with God. Remember the story of the Prodigal Son; keep in mind the Lost Sheep. God is our refuge and our security."

Because of this relationship in Christ, our friendship grew, grew in leaps and bounds. In the meantime, however, once every fortnight or so, he got so blue and despondent that he wanted to give up the struggle. Three times he attempted to end his own life. Each time I had to spend hours comforting him and encouraging him by reminding him of the sufferings and the spirit of our Lord.

One night, not many days later, while we were lying side by side, he confessed and confided his secret to me. He was in the "special goods" (opium) business, and his branch office in Kunming was to collect the special goods in Yunnan and ship them to Hongkong. In the corner of his mind he was constantly troubled and agonized by the thought that he was imprisoned for this evil business but, in fact, he was saved by it because, utilizing his connections in this special goods business, he was released two days earlier than I.

After he had heard of my release, he paid me a visit and told me that he was required to redeem himself by doing the same business for the Communist government. Before his departure, I told him frankly, "You are a Christian. God has saved you a number of times, when you attempted to end your own life. Listen to God rather than to man."

"Don't worry, Old Bishop! I am a Christian now, though not before. I have found my little faith through my own experiences. I'll keep it and treasure it!"

4

Life In the Wooden Cage

I WAS PUSHED and literally "packed" into the wooden cage. It consisted of wooden bars from four to five inches in diameter each and about two inches apart. The door was closed, locked, and chained. The cage was about six by eight feet in dimension, situated in a dark, damp end of a long room. The other end of the room was used as quarters for the guards who watched us inside the cage and also those outside who were waiting for the judgment of the Commissioner of Police. Outside the wooden cage a pile of ropes, chains, and handcuffs was stored and ever ready for use on prisoners.

In this small wooden cage I found (as I later counted) 18 new companions—criminals of all classes, including thieves, robbers, murderers, counterfeiters, adulterers and, of course, the so-called "political criminals," including myself. We were packed in like sardines, with no space for lying down or sitting; consequently, we all had to stand. It was fortunate that this all happened in winter as we did not mind so much being packed together, because we had the one benefit of keeping ourselves warm in the cold weather.

When arrested, no one was able usually even to say good-bye to his loved ones or to bring any necessary things, except what he wore. Those who in due time became tired, exhausted, or sleepy simply slipped down to sleep as best they could, with the legs and feet of fellow prisoners all around. This was what we called "slumber among wooden and flesh bars." While the ground space was thus taken by the fast sleepers, the rest had to doze against the wooden bars and one another.

The Communist regime provided neither water nor food for the prisoners at the police stations. Either you had to pay the guards heavily to buy these things or you were provided with them by your family or friends through the guards, whom you were obliged to tip for the favor of bringing things in for you. Furthermore, where the prisoners were kept was unknown to their family members unless messages could be sent by means of the almighty dollar—through the guards again. At the same time, wild rumors were circulated concerning the political criminals; so frightening that the timid, though close and intimate, friends and relatives were afraid to call on the prisoners either at prisons or police stations. Consequently, the majority of the political criminals were left without food or water for days, and had to depend on the mercy of fellow prisoners imprisoned for other offenses, who were allowed to see their relatives and friends and have their necessities sent in directly twice a day.

Conditions were the same in regard to our daily trip to the latrine. The prisoners with money were attended to punctually; those without it were neglected or much delayed. These irregularities, neglect and delay, particularly with diarrhea cases, soon turned a part of the wooden cage into a natural latrine, filthy and foul smelling—with lice, fleas, and hungry rats, both big and small, "geegerling" and running hither and thither even in the daytime, from the farther corners of the muddy walls just next to the wooden cage.

At 12:20 a.m., I was called by Judge Yeh of the Police Station to go to his bedroom-office for registration. Though a native of Yunnan, Mr. Yeh did not have the Yunnan provincialism, as he had served as judge for years at various courts along the coasts and the Yangtze Valley. After my registration was completed, we two talked congenially for about an hour about each other's past life and stories.

To express his deep concern and sympathy for me, he said, "Today I am working as a judge, an official, here, but tomorrow I may be one of your companions in the wooden cage, too. In order to redeem our past and hold on to our present positions, we, the former government workers—not members of the Communist Party but working temporarily for the new regime—are required to report or arrest at least twenty political criminals or suspected persons. Some utilize this opportunity for personal revenge and others for getting credits for their future career. Society is certainly in a mess —a reign of terror! Only the people are suffering. As far as I know, at least ninety per cent of the people arrested these days are innocent. We have been ordered to keep you here but we have no authority either to give you a trial or to release you, even when we know you are innocent.

"All the authority is in the hands of those working at the Public Safety Bureau, the new Communist organization, from which even Governor Lu Han is getting his orders now. The head of that organization is Mr. Sung Yi Heng, formerly trained in Yenan, Shensi [the Red capital before the Communists took over the government], who has worked for years in the past as a small clerk of the provincial government. No one would suspect him as a Communist underground agent, but he is, and he has been appointed by the new regime in Peiping as temporary head of Yunnan here.

See, all arrest orders are issued in his name; nobody else has any power.

"If you have any strings to pull, pull them quickly, I advise you. As an example, Mr. Chang, head of the Southwest Public Highway Bureau, was arrested and kept in the wooden cage for three days but, by pulling some strings at the Public Safety Bureau, he was released yesterday by Mr. Sung's order, even without a trial."

"How can he do it?" I inquired. "According to law, the police authorities have no right to detain a person in the police station for more than 24 hours. Furthermore, you cannot arrest or release a person at will. Whether a person is guilty or innocent, he is entitled to have a trial. I expect to have a trial before you—as soon as possible."

"Oh, Bishop, you are too innocent! The old law is cast overboard and gone! If the laws of the Nationalist government were to be enforced, my position as a judge would be 'as firm as the Tai Mountain,' " he exclaimed, and continued, "but this is the time of revolution. There are bound to be lots of injustices and scapegoats, especially when the new regime believes in the principle that 'it is better to have 10,000 innocent imprisoned than to let one guilty go free.' "

"Friend," I asked again, with a deep sigh, "then am I going to be in this wooden cage, packed in like a sardine, for days and months, suffering both mentally and physically, for no crime and without even a trial?"

"Don't worry," he said, trying to comfort me; "I know who you are and I'll do my best to help you if I can, by getting you transferred to the best prison in the city, the district jail, where you may find life comparatively easy."

On my return from Judge Yeh's office, I was horrified to see, for the first time in my life, a girl about twelve years old, with hands and feet tied, pulled up by a rope about seven feet from the ground under a big leafless tree in the courtyard. Her mother, a woman about forty-five years old, lay against the tree, crying and screaming. Two policemen, with long bamboo sticks in their hands, beat the girl alternately with all their might, trying to force her to tell where her father was. (He was supposed to be one of the great political criminals.) The girl was yelling shrilly, "Help! Save my life!" For twenty minutes the beating went on incessantly, without the slightest mercy.

After I was shoved into the wooden cage again, I told our fellow prisoners what I had just seen. We all became indignant and considered that the policemen were outrageous for beating up a young girl like that. Then our Christian friend, Mr. Hsiao, asked the guard standing at the door and watching the torture, "What crime has that young girl committed? It is terrible to torture a kid like that!!"

The guard, seemingly a little tired of looking at the screaming girl, turned his head toward us and said, "That girl certainly has a tight mouth. We have tried all sorts of ways and means to get her to tell us where her father is, but her answer is always, 'I don't know.' You know, we have been ordered by our new boss to have her father arrested today but we have no clue as to where he is. There is no fooling with any order from the new regime. If we get an assignment and fail to fulfill it, we will be tortured before we are fired. That girl knows where he is; she just won't tell!" Before the conclusion of our conversation, both mother and child were brought into the room, tied with ropes, and laid on the muddy floor just outside our wooden cage. They were murmuring and groaning in agony!

Early at night, three of our fellow prisoners were taken out of the wooden cage for a trial, so I got a little extra space and sat down with my head resting on my knees. Soon I went to sleep—sound asleep—after the strenuous mental and physical strain of the past two days.

But at five o'clock in the morning. another incident gave me a second shock. I was awakened by the clicking of chains and pounding on the big Chinese iron lock, and the stern voice of a policeman ordering us in the cage to make room for three victims, pale and motionless, being carried in one after another. According to information circulated among the old timers, they were two carpenters and one apprentice, natives of Hunan province. During World War II they had walked to Kunming as refugees. They had worked hard ever since and prospered in their business, a carpenter shop on Ching Yun Street.

Behind the carpenter shop, there was a big modern mansion belonging to a committeeman of the Yunan provincial government. Soon after the surrender of the Governor, the owner was thrown into prison and the mansion occupied by an important official of the new regime. Four nights before, the mansion was visited by a group of thieves and some valuables taken away. The new neighbor became enraged. Suspecting the three carpenters, he ordered the police to have them arrested. They were first threatened and then beaten with clubs, yet they would not confess that they had done the stealing.

At last an implement, called in Chinese, "Tien Hua Chia," was used to torture them in order to get the necessary confessions. This is a crude electric machine made of a small wooden box in which a battery is stored, with a wheel, and two cords which are attached to the prisoner's hands or some other parts of his body. When the wheel is turned, the victim receives electric shocks. Ordinarily, any person who has gone through that sort of torture is ready to make any confession as told, but with these three Hunan carpenters this torture lost its effect. They had no confession to make and they didn't make any, although they were half dead, motionless, and speechless, lying there with us.

After that, my strong belief in Mencius' theory that "human nature is good" began to shake in my mind, and I felt that Hsuan Tze's conception that "human nature is evil" might have some truth, too. I became quite a skeptic! However, I was not skeptical enough to question the sincerity of the Communists. Before they took over the mainland, they had talked and propagandized so much about personal rights. I thought, "What right do they have to imprison so many of us without trials and to torture the carpenters without evidence? What kind of democracy is Communism in China, as they call it the 'New Democracy?'"

We would hardly distinguish between day and night in the dark end of the long room, except for the light over the guards' desk at the other end, which made night brighter than day. Time for us in the cage was one monotonous anxiety. We had no water to wash with, no pen to write with, nothing to do—just wait and hope for an early trial. However, there was *one* change—the change from standing to squatting, or vice versa!

In spite of the cold winter, we seemed to have no need of any more clothes than we had on. Undoubtedly we did keep ourselves warm by being packed together; in fact, sometimes it became too warm for lack of fresh air and light. Periodically we were suffocated by bad smells, and could not help vomiting at times. Because of our hunger and thirst, isolation from our loved ones, congestion with our fellow victims, the filthiness and bad smells, our helplessness and anxiety, we were indeed suffering both mentally and physically. On several occasions, Mr. Hsiao, our Christian companion, could bear no more and became restless and very much disgusted. More than once I commented, "This is real hell on earth."

Only in suffering and distress did I begin to realize human weakness. Not until we have lost freedom do we appreciate freedom; not until we have been confined and isolated do we crave fellowship with others; not until we are face to face with human sufferings and have gone through them ourselves do we see some value in the happiness of life, the blessings of God! Though I was constantly haunted by fear that I would not be able to "endure to the end" and that I might bring shame upon God's name, I was at least grateful to God for all the past blessings of life!

Twenty-four hours were gone! There was no sign of an early trial which might lead to release. Intuitively I felt that we were heading for a long confinement, but instinctively I refused to accept our lot, and hoped and prayed for early restoration of freedom. The first twenty-four hours seemed like twenty-four months or years.

While I was in agony and bowed my head, praying to God for strength to endure, an officer dressed in army uniform, whom I had never seen or met before, suddenly stood outside the wooden cage asking for me. He introduced himself, whispering to me, "I am brigade commander of the newly organized Communist Army and a relative of your good friend, Mr. Chang [fictitious name]. All your friends and Church workers are very anxious about you but afraid to come themselves. I am their representative to see and help you. I am not a Christian but we appreciate what you and your Church have been doing for Yunnan. Don't worry; have patience just for a few days. You will be out with us!" After his visit, he went in to see the authorities at the police station and later, out to see those at the Public Safety Bureau, the highest authority in Kunming at that time.

For the first twenty-four hours I did not have any food or water. I was hungry, thirsty, downhearted, and helpless, and had begun to wonder whether or not I was forsaken by both God and all my friends and colleagues. But that visit of the brigade commander gave me unlimited comfort and hope. Up went my spirit! After his visit, I was asked by our Christian companion, Mr. Hsiao, to share some of his food which he got through by bribing the guards with five silver dollars. I helped myself and shared with him my encouragement and hope.

Next morning, instead of another personal visit by this God-sent helper, a written message came from him, saying, "The whole situation is not good. The present policy is to arrest, not to release. Be patient, Bishop; we'll do our best outside!" This message, though pouring a great deal of cold water on my expectations, did not kill my hope. It further confirmed the wild rumor that the Communists during the first ten days of their control—beginning from the withdrawal of the Nationalist Army from Kunming on December 21—had arrested close to ten thousand civilians in the city of Kunming. Arrest! No trial! No release! A way of God, surely, to teach us patience!

On the third morning, through the efficacy of four more Yunnan silver dollars from Mr. Hsiao, the guard, formerly very stern and fierce toward us, became quite merciful and provided Mr. Hsiao and me with a basin of cold water. At our repeated requests, he let us out of the wooden cage, and we washed our faces and hands alternately. Soon the water became black and muddy, yet we enjoyed our first cleaning and felt greatly refreshed. Simultaneously, our food was brought in.

Through much begging, another courtesy was granted in that we were allowed ten minutes to eat our meal, standing at one end of the desk used by the guards. It was indeed a valuable ten minutes and a great treat! Standing by the desk, we could through the door see the blue sky once again, tell the color of the food we were eating, stretch our arms, bend our knees, and kick our legs and feet without bumping against our neighbors. Our food seemed more tasty and the air fresher. In a barbarous manner, we swallowed down the rice and other food in no time. The guard, in spite of his merciful grant of ten minutes, repeatedly told us, "Hurry up; the inspector may come at any moment." At the end of eight minutes, he pushed us into the wooden cage and we were locked up again.

5

Parade of Prisoners

NO SOONER were we pushed into the wooden cage than we heard four soldiers, fully armed, marching into the courtyard of the Fourth Police Station, with a commanding officer behind. Entering the long room and handing a piece of yellow paper to the guards, the commanding officer yelled, "Here is the order. Get those four criminals ready quickly, for we must go immediately!"

The door of the wooden cage swung open again and four of us, including Mr. Hsiao and me, were ordered to come out. Where we were being taken, no one would tell, and if it were for better or for worse we had not the slightest idea. Glancing over the order put on the guards' desk, I found just four names with the same vague title of "suspected political criminals." That title did not give us any clue as to the charges against us. At a time like that, it might mean anything from the least important to the most serious case. However, it was the first time we learned something about the accusation against us.

After we came into the room, one policeman soon gathered up a pile of ropes, another a pile of leg shackles and four pairs of handcuffs. Two of them came over and attempted to tie the four of us first with the ropes. Immediately came an indignant uproar from us! One jumped up and scolded them, another argued hotly, the third nearly hit them with his fists, while I, the fourth, said to them rather courteously, "We are neither thieves nor robbers. For days we have waited for trials. You cannot treat us like this, with leg chains, handcuffs and ropes, before we are found guilty and sentenced. Why don't you pack us in a truck or cart and send us to our destination?"

"No," one of them answered reluctantly, "we have been ordered to parade you on the streets first, you four—for what purpose, I don't know."

"All right, we shall be in the parade and go anywhere you lead us, but what is the use of tying us with ropes and putting leg chains and handcuffs on us?" I asked.

"If any one of you should run away, what would we do?" exclaimed another policeman.

"That is simple!" answered Mr. Hsiao. "You have guns and pistols, while we are barehanded. If you see any one of us trying to run away, shoot him!" The rest of us chorused, "That is right! Shoot him! Kill him!"

In the midst of our hot argument, someone exclaimed, "Is this the way people are handled by the new regime? Are we not the people?"

"No, absolutely no, you are not the people," the commanding officer shouted in a stern voice. "You are only prisoners and reactionaries. This is the order which we know only to obey." Turning to the policemen, he continued, "Hurry up! Tie them up! We have to go and complete our mission!"

"What are our crimes? We have never been tried or sentenced," we shouted.

In the midst of the uproar and turmoil, Judge Yeh came up with a compromise by proposing that ropes be used to tie our upper bodies only and that our hands be handcuffed two by two, saying he would be responsible. He believed it would be enough and that none of us would run away—"so let the leg chains go." The soldiers and policemen were getting impatient and becoming more stubborn every minute, and I instantly recalled a common saying among us Chinese, "When a scholar meets a soldier, reason has no room." I suggested that we accept the compromise and became the first victim to be tied, with my right hand behind my body and my left hand handcuffed to the right hand of Mr. Hsiao.

Putting down the sleeves of our top coats to cover our handcuffs, Mr. Hsiao and I walked in front as twin brothers. To him, being a baptized Christian, I said, "In Christ, we Christians should be bound like this, both in distress and happiness!" He replied jokingly, "I wish I could take a picture of myself being handcuffed with a bishop, as criminals of the new regime!"

A few minutes later we were ready to start our parade. A policeman, with a gun on his right shoulder, led the way and we four "criminals" followed, walking in two rows. Two soldiers walked on either side and the commanding officer behind, all having their pistols pointed at us and their fingers on the triggers, ready to fire at any moment.

Slowly we marched on, first through small alleys and then into the big main street of Kunming, called Tseng Yi Road (Road of Righteousness). As soon as we stepped into the Tseng Yi Road, which we had known so well for shopping purposes, I was reminded of the Chinese philosophy of Tseng Chi, or Spirit of Fidelity. So I bowed my head and prayed in my mind, "O God, give me Tseng Chi [fidelity] to fight for Tseng Yi [righteousness]!"

It was a terrible march of disgrace and humiliation! Step by step we walked on. We seemed, in this parade, to have something that attracted people, drawing them out to see us, yet, at the same time, some impulse impelled them not to speak to us. Friends and relatives of the group turned their heads away and pretended not to have seen us. Here and there among the crowd we heard some people telling others, "They are political criminals being taken out to be shot."

Whither we marched, neither the soldiers nor the policemen would tell us; what the fate ahead of us might be, we could not tell. In my mind I was wondering, "Are we marching down the Tseng Yi Road to the Flower Circle outside the South Gate? Are we going to be shot there without a trial?"

If I were to be shot in the Flower Circle, I would plead to be shot in our nearby cathedral—the Allied War Memorial Church built in memory of all the Allied forces who paid their highest tribute in World War II. So again I prayed, "O God, Thou knowest I am not a political criminal. If I am to die, I beseech Thee to let me die as a martyr of faith in Thy house, which is so near. Move them to grant me this plea."

While my mind was wandering, questioning, and praying, the policeman leading the procession suddenly looked back at us just before we reached the South Gate and slowly turned into the left street, marching on toward the East Gate. We knew the district jail was located along this street. Farther on, to the outside of the East Gate, there was a big public playground where many criminals had been executed. "Are we going to stop at the district jail for our trials or proceed on to the public playground to be executed?" were the immediate questions rambling in my mind.

Parading on the streets, in shame and humiliation, the four of us had had our eyes constantly on the pavement, except only occasionally when we looked up trying to find out where we were. We plodded along the stone pavements just like sheep being led to the slaughterhouse. Suddenly, the guiding policeman stopped. Raising my head, I saw the red Forbidden Gate, above which a sign was clearly written in Chinese, "The District Jail of Kunming."

The parade was over and at one o'clock in the afternoon we arrived at our destination—not death, but jail! Hope was high, for in jail we might have our trials. Conscious of having committed no crime, I was grateful for this transfer and expected an early release. In my mind I said the same thing St. Paul said, because I knew I was "persecuted but not forsaken." "Marvel not, brethren, if the world hateth you . . ." If you cannot "bless them who persecute you, curse not," and at least, we should "pray for them that persecute you."

Praise be to God! In my distress and agony I *did* pray for them, beseeching God to move their hearts that they might mean what they said—freedom of religion—and that "we . . . . by Thy mighty aid, might be defended and comforted in all dangers and adversities."

6

God's Providence and My Religious Work

STOPPING at the big Forbidden Gate, we four were surprised to see that the gate had two big doors swinging to either side. They were newly painted in Chinese red —so bright and striking! Those two red doors were always locked and barred, but on the right there was a small door, which was opened only to let new prisoners in. The Forbidden Gate was guarded day and night by soldiers, fully armed. It was "the gulf fixed"! Neither relatives nor friends could pass from without, nor could the prisoners pass from within. "Look at the red doors, so bright and clean," I said to my fellow prisoners as we waited together. "Today we are the prisoners of the *Red*. I just wonder whether it is as clean and bright inside!"

"Don't worry, you will soon find out," said Mr. Hsiao.

At the Forbidden Gate of the district jail stood Mr. Ming, the first secretary to the warden, checking and registering all new arrivals. Later I learned that the warden, a native of Hupeh who belonged to the old school and was not a Communist, had been cast overboard while Mr. Ming, as a candidate for full membership in the Communist Party, had been appointed by the authorities of the new regime to take his place and was entrusted with the management and investigation of the prisoners. He was a young fellow about twenty-five years old, and actually had complete authority in dealing with the prisoners.

I looked at him but could not recognize him. While I was talking he saw me and after looking at me for a few minutes, he came over and asked, "Aren't you Bishop Huang?"

"Yes," I replied. "How do you know me?"

"Don't you remember me?" Mr. Ming asked, without waiting for the completion of my question.

"No, I'm afraid I don't remember you at all, because these days my eyesight is no good and neither is my memory. You will have to excuse me, please," I apologized.

"Well, I attended your lectures at Chung Hua Sheng Kung Hui [Holy Catholic Church in China] in Kutsing when you were there last year," explained Mr. Ming. (Kutsing is a small town east of Kunming where we had a small hospital and a church. It was my custom on an episcopal tour to stay in each station for a few days and give a series of lectures on Christianity to both Christians and non-Christians. Evidently he was one of the non-Christians who had attended my lectures in Kutsing.) "How do you get here? For what?" he continued.

Happy and excited to meet him and to have his unexpected friendship, I answered, "By the providence of God I am here and meet you again. I don't know the charges against me or my accusers. We haven't had a trial yet and I guess it is a God-given opportunity for us to learn some lessons of patience!"

Trying to comfort me, he threw up his hands and said, "Well, never mind, I'll take care of you." Turning around to one of his subordinates, named Wong, he ordered, "Send Bishop Huang and Mr. Hsiao up to Room 3 West and tell Mr. Chen to take good care of them." Praise and thanks be to God for His special grace in giving me Mr. Ming's warm friendship. Through him I was able to get into Room 3 West and have a number of privileges.

The jail was a very small one, built in a U-shape, with three two-story buildings, the main building in the middle and two small wings on either side. The main building had twenty-four cells, twelve on each floor, and the two wings had altogether twelve rooms. Except for three rooms on the second floor of the west wing which were used as "privileged" cells, the other nine rooms were used for an inner office and bedrooms of two Communist leaders, a coöperative store, clinic, kitchen, and the living quarters of watchmen and guards.

There were altogether twenty-seven cells of standard size, 9 by 12 feet. Each cell had a small window, close to the ceiling, two feet by one and a half feet in size, with ten iron bars. Joining the two wings in the front there was a solid high brick wall, with a small courtyard, 30 by 40 feet, in the middle. Outside the brick wall there were three or four small rooms and one big hall attached, used later by the Communists to hold trials.

In the open alley there were half a dozen small holes dug in the ground, each about three or four feet deep, one foot wide and two feet long, which were used as latrines. Twice a day, at 7:30 a.m. and 4:00 p.m., all the prisoners were lined up under guard to take their turns at the latrines.

In the middle courtyard, 30 by 40 feet, 407 prisoners, consisting of 317 political criminals and 90 others, might have their daily exercise. By exercise, they meant standing in the courtyard and breathing some fresh air. When the cells were unlocked, the courtyard was jammed with people; we could hardly find space enough to stretch our arms and kick our legs. This was also the place where prisoners might buy hot water to drink for ten cents a cup (about two and one-half cents U.S.), and get cold water once a day for washing faces, etc. This was limited to one basin each. Because of poor drainage, half of the courtyard was flooded with filthy water most of the time. This was also the "Times Square," where news and rumors were spread like wildfire by whispering one to another.

Following Mr. Wong and entering the privileged cell, I was properly introduced to Mr. Chen, the head of the cell (formerly called the "Dragon's Head," which name was forbidden to be used by the Communists), and I was happy to find that both Mr. Chen and I came from the same province, Anhwei, in the Yangtze valley. With the request from the first secretary, Mr. Ming, and this provincial tie between us, Mr. Chen immediately took special interest in me and asked me to help myself to some of his cookies and tangerines. By his order, also, I was exempt from participating in their daily gambling game, on the ground that I was a bishop in the church and, naturally, did not know how to play. However, I was required to pay my share of ninety dollars Yunnan silver ($22.50 U.S.) for three privileges in the cell: the limitation of prisoners to twelve persons in one cell (while in other cells the number might go up to twenty); the privilege of boiling our own drinking water over a charcoal stove; and that of climbing up to the high small window in the cell that we might have a look between the bars at our friends and relatives standing outside the jail and talk to them in the "deaf and dumb" signs of our own invention, if we were lucky enough to time our meetings.

I also had to serve the oldtimers a little. According to the unwritten law of the jail, a newcomer was to obey the leader in the cell, usually occupied the worst place near the door, and did the dirtiest work—unless he had lots of money to buy service from others instead. Mr. Hsiao and I were the newcomers, while the other nine in the cell had come about one week earlier. He had seven dollars while I had only three, and both of us were cut off from our sources temporarily. It was very kind of Mr. Chen, in spite of our poverty, to require us only to sweep the floor three or four times a day, clean the chopsticks at mealtime, roll up the bedding in the morning and spread it out at night. It was not much and we were glad to do it. In addition to these requirements, I cleaned up the whole cell, with Mr. Hsiao's assistance, and washed the floor on the first day. Thus it was made rather comfortable and everybody was happy.

While I was doing that, I told them, "We Christians are glad to serve others in any way possible, as our Lord said, 'I am come, not to be ministered unto but to minister.'" Curiosity and interest were aroused, particularly in Mr. Chen, who, immediately stopping his card game with the others, requested, "Tell us something about your Lord. I am surprised that you, as the Bishop of Southwest China and a big shot, are so happy and willing to render service to others. I am one of the two leaders of the Ching Pong in Yunnan." ("Ching Pong" means Green Secret Society, which was originally organized at the beginning of the Ching Dynasty for the purpose of restoring the Ming Dynasty. It has lost its original purpose now but still remains very influential among the middle and lower classes of people. The members have their secret signs and talks, and consider themselves as blood-covenant brothers. Fidelity to the Society is their cardinal virtue, and mutual assistance their bounden duty.) "As the head, I am always served; what I do is to give orders and everything is done by my subordinates. I have read once or twice about your Christian service to others, but I have always thought that it was mere talk. I never realized that you actually do it. Now, come on and tell us about your Jesus."

Gripping the God-given chance, I started my preaching to a group of six cellmates, two of whom showed intense interest, while the other four were rather indifferent. I gave them some highlights on the life and teaching of Jesus Christ, particularly His love, service, sufferings, and death. After my talk to the group, we had our first meal together in prison, and Mr. Chen insisted that I should tell him some more personally about Jesus, so we two spent our first evening in a corner by ourselves. I did the talking while he listened and questioned me. For three hours I talked, and told him how Jesus served His disciples by washing their feet, and healed the sick, and how He was accused, tried, and condemned to die on the Cross, in order to live up to what He had preached on love and sacrifice. I grew tired of talking but his interest increased as time went on.

Next morning about ten o'clock, by his self-assumed leadership of the three privileged cells, he got a few more fellow-prisoners to attend my lecture. This time, I spent one whole hour on the sufferings of Christ and St. Paul, intending to give the gloomy prisoners some consolation.

On the second evening, one of Secretary Ming's assistants, deeply depressed, came into our cell and told us about his sufferings because of gonorrhea. On the side, I casually commented, "That is your punishment, boy! Why don't you go to some hospital and see a doctor?"

"That costs money, and I don't have it," he answered.

After thinking for a moment, I said to him, "I think I can help you, if you are willing to do me a favor. I have a bottle of sulfadiazine and a bottle of soda mint at home and with those two bottles of medicine, I think we can do the trick."

"That is just the medicine I need. Nowadays it costs one dollar [Yunnan silver] a tablet—too much—and I cannot afford it," he shouted in rapture. "What is the favor I may do for you, sir?"

I answered, "I shall write a note. With that note, you go yourself tonight to the Diocesan Center in the Ching Kuo New Village to see a colleague of mine. From him you will get those two bottles of medicine and, in addition, I want you to get sixty copies of the New Testament in Chinese from him and bring them back with you. That is the favor. Are you willing to do it?"

He hesitated a little, then asked, "Sixty copies or six copies?"

"Yes, sixty, not six," I answered.

"Well, that is too bulky. I cannot hide sixty copies of the New Testament." After a while he continued, "Let me go and try to get permission from Secretary Ming. I'll do my best to get it from him. Now you write the note." Out he ran. Ten minutes later he hopped and jumped back, and said, "Give me the note. I am going over there right now." Two hours later he came back and walked upstairs. Unlocking our cell, he appeared very happy and handed me all the things I had asked for. I began to treat him and in a week he was much relieved.

Next morning we organized our first Bible class and each member had a copy of the New Testament. By the beginning of the second week, we had three Bible classes, one in each privileged cell. Each Bible class had one hour of Bible study on Mondays, Wednesdays, and Fridays. On Tuesdays, Thursdays, and Saturdays we spent one hour each in discussing our

personal and religious problems. Thank God, I was busy
again with His work every morning.

Soon "Old Bishop" became my name. By adding the adjective "OLD" to my title, my fellow prisoners were trying
to pay me some respect. (By "Old" we Chinese mean "full
of wisdom and experience.") Also, my manual duties were
taken over by the younger members of the Bible classes. In
our meetings, I could see and feel God's presence with us
and His working and moving power over those non-Christians, including two Communist agents (as I later discovered) who were imprisoned with us. Without God, it
would have been absolutely impossible. They were just
hungry and thirsty for Truth—Truth made them free, free
from fear and worry. They were ready to face sufferings
and to forgive injustices done to them. I myself was overwhelmed with gratitude to God for those happy days in
prison with them.

By the end of the second week, some members of the Bible
classes and a few Christians, including Mr. Feng, Mr. Chu,
and Mr. Lin, suggested that we should start Sunday services
in jail. We worked hard on this proposal and pulled every
string possible, but we failed. Meetings on Sundays were
absolutely forbidden and any books other than those on New
Democracy and Communism were not allowed to be brought
into prison.

To get around that, I invited all the Christians to the Bible
classes and discussion meetings on the weekdays. We started
our meetings always by reading the Lord's Prayer together,
and concluded them with a few simple prayers for the country, the Church, justice, patience, sufferings, and our loved
ones. The most welcome and touching prayers were those
for the particular members of the classes and their beloved

ones. Such prayers usually aroused a great deal of emotion, sympathy, gratitude, and sometimes even weeping. Instead of Sunday services, we had daily services. Often in my prayers I said, "O Lord, we praise Thee and thank Thee, for 'very worthy deeds are done . . . by Thy providence!'"

7

Picked as an Example of "The Enemy" to Be Reported

BY MEANS of the Bible classes and discussion meetings, held daily except Sundays, intimacy was cultivated among the members of the three privileged cells, together with a few Christians from other cells. In Christ, fear was discarded, sympathy was developed for one another, and stories after stories of how and when and where they were arrested and imprisoned were told.

Putting all the stories together—the outcome of an intimacy which no superficial observer, either foreign or Chinese, could have—we were able to gain a deep insight into the first stage of the Chinese Communist regime, the so-called Establishment of the Revolutionary Order. Beginning from December 20, 1949, after the agreements with the Nationalist armies were carried out (including the release of the National Army generals, a gift of a few truckloads of gold and silver, and the withdrawal of the Nationalist forces from Kunming), the Communist underground workers, together with their superpolitical organization called the Public Safety Bureau, came out and took over complete control of Kunming. Immediately, five wooden boxes were set up in five different sections of the city. By means of newspapers, handbills, posters, group meetings,

radio, etc., the masses of the people were told and encouraged
to report secretly through the reporting boxes, the names
and addresses of secret agents, spies, reactionaries, counter-
revolutionaries, imperialists, feudalists, and all the enemies
of the new regime. The people, as they were told, were now
the masters of the government of the people; it was their
duty to report such persons in order to show their loyalty
and help establish the Revolutionary Order. Each of the
former government employees was required to report at
least twenty such persons in order to redeem his sinful past
and to hold his position in the new government.

The wooden boxes were soon flooded with names, addresses, and accusations. Without any investigations whatsoever of evidence or reports, whether true or false, the first
and only job then of the Public Safety Bureau was to arrest
and imprison all the accused. It was the same old policy of
the Communists: "Better to have ten thousand innocent imprisoned than to let one guilty escape."

Simultaneously, the Communist and radical students and
youngsters were organized into a "Self-Defense Corps,"
armed with rifles, and authorized with unlimited power to
arrest any persons whom they did not like or thought reactionary.

Over night, unions of servants and workers of institutions,
both government and private, came into power and took over
their administrations and sent the former administrators to
jail.

Teachers and professors who had been strict and rigid
administrators were hard hit. In our jail we found the superintendent of the water works, the head of the railroads, man-
agers of banks, editors of newspapers, head of the overseas
bureau, principals, teachers, deans of schools, professors of
colleges, the head of the highway bureau, priests of the

Roman Catholic Church, and a Protestant bishop. There were also monks of other religions, elders of secret societies, radio engineers and superintendents, heads of publishing companies, doctors, lawyers, and what not. It was indeed a gathering of a variety of talents.

In a few days all the old jails and many newly set up prisons in Kunming were jammed with political criminals. According to the lowest estimate, more than 7,000 civilians were arrested in Kunming within a week, excluding several thousand imprisoned military policemen and soldiers of the Nationalist government. It had always been considered disgraceful to be imprisoned but, during the reign of terror, so many leaders and educated people were imprisoned that we could often hear such comments repeated as, "It is no more a shame but an honor to be in jail; it proves that we have been conscientious and strict in our works."

Some people were arrested without charges, others without accusers, and many times, if the names and addresses of informers or reporters were given, they were found upon investigation to be non-existent. As long as your name and address were dropped into one of the secret reporting boxes, or any one of the Self-Defense Corps or the unions disliked you, your fate of being imprisoned was sealed.

This was the so-called first step, that of purging the enemies of the people and the establishment of the Revolutionary Order. This period was also called the period of arrest and lasted from four to eight weeks, during which nightly curfew from 8:00 p.m. to 6:00 a.m. was strictly enforced. Arresting teams, consisting of youngsters (members of the Self-Defense Corps), policemen, soldiers, and others, led, usually, by the members of the Public Safety Bureau, went from house to house between 12 and 5 at night, searching the dwellings thoroughly and picking up the accused quietly. It made no difference to them whether old or young, male or female, educated or illiterate, high-ups or downtrodden, rich or poor, innocent or guilty. They were all hushed up and shoved into prisons for indoctrination, confession, confiscation, utilization, or liquidation, as the Communists saw fit.

After the Communists had taken over the city, some newspapers were suspended from publication and others were seized and utilized as agencies of Communist propaganda. There was no news, either national or international, but they were full of articles on "Learning," "Increase of Production," "Communist Songs," "Heroes and Heroines of Labor," "Minutes of Students' Meetings," and particularly accusations against those leaders in society who seemed prominent, good, and dignified. This was to make the people alarmed and aware of the enemies of the people, so as to report them during the period of establishment of the Revolutionary Order. Only in the case of those picked out to be examples as enemies of the new regime were the charges made known to the public as well as to the accused. Ordinarily no charges would be told and the accused were required to confess them in their "Confession Papers." (This will be dealt with in detail later.) I was rather fortunate, in a way, to be picked out as one of the examples in Kunming.

On January 5, 1950, there was a long article, with big headlines, about my case published in the Ping Ming Daily ("Ping" means common; "Ming," people). It was an old paper taken over by the Communists who kept the same name and sent three former editors and half a dozen staff members into the same jail with us. As the Communists published no news, and the prisoners were getting tighter and tighter financially, the people lost their desire to read the daily papers and soon only three or four copies of the Ping Ming Daily were bought in our jail. In the daytime I heard two prisoners whispering to others, "That is the bishop," after they had read the article. They seemed to think that I had actually committed the worst kind of crime and that I was really a great criminal. Not until late in the evening did one of my cellmates whisper to me, "Have you read the article in the Ping Ming Daily today about yourself?"

"No, I have not read it. Do you have the paper? I want to read it," I said. After a while I continued, "That is why some prisoners looked rather strangely at me, and I heard two of them saying at different times, 'That is the bishop.'"

"Oh, it was a long article," he said, "and it published so many charges against you I can't remember them now, but I do remember it was written by an organization called the Christian Fellowship in Kunming. By the way, your friend and student, Mr. Li, from the Great China University, has a copy. You may get it from him and read it yourself tomorrow."

For the whole night I could not sleep, thinking and pondering over what they could have accused me of and praying for guidance and wisdom to refute them. I was nervous and very anxious, and wished that daybreak would come sooner and the cell door would be unlocked earlier so that I might jump up and rush to Mr. Li's cell to get that paper. In the midst of my sleeplessness and anxiety, there was one consolation. I was sure there had never been an organization called the Kunming Christian Fellowship, although we had a united project named the Association of the Christian Organizations in Kunming. I had never heard of this Christian Fellowship. If there were such an organization, I could bring a lawsuit against the paper for the publication of that article and force my accusers to take an open stand against me. It was all such naive thinking on my part! The law had been declared nil; judges were gone. In their places only teen-age youngsters held trials and gave sentences according to the Communist yardstick—everything decided and done for the benefit of the Communist Party-State. There was no justice, in the sense of our understanding of the word.

Early next morning I got up and dressed, and waited. As soon as the cell was unlocked by the guard, I rushed to Mr. Li's cell only to find that he had given the paper to somebody else; however, he promised to get it back. I immediately ran to other cells for the same purpose. Friends helped, but we could not find a single copy.

Two hours later Mr. Li came to our cell, expressing his regret that his friend and a few others had used it as toilet paper this morning. However, he whispered to me from memory the charges against me, which were confirmed later. First, under the cloak of religion, I had spread my spy network in both Yunnan and Kweichow provinces; secondly, I was a spy of the Nationalist government sent to Kunming to check the activities of the Communist underground workers, especially those of Governor Lu Han; thirdly, with funds of the Hueitien Hospital (our Church hospital) I had sent Church workers to the States to be trained as spies against the "government of the people"; fourthly, I was one of the greatest imperialists, giving orders all the time to my subordinates; and lastly but not the least, I was the Number One "running dog" of American imperialism and Number One American spy in Southwest China. The article was signed by the Kunming Christian Fellowship.

With those concrete charges and the accuser in mind, I

was able to explain things to my friends and cellmates in prison. Time and again I brought the charges up in my confession papers and expressed my willingness to pay the extreme penality if they could be verified and proved. Without knowing me or the policy and work of the Church, and having neither time nor chance for investigation, the masses of people would take it for granted that I was a big spy, under the cloak of Christianity, and, naturally, they would become more suspicious of the leaders of society, no matter how good they had been. As a result, they would report secretly on the slightest suspicion, in order to safeguard themselves and also prove their loyalty to the new regime.

8

Communist Blackmail

ON JANUARY 8, 1950, three days after publication of
the five accusations against me in the Ping Ming
Daily, and after my mind had begun to settle down a bit,
startling news came from our Diocesan House. Two young
fellows in plain clothes, who claimed to be officially in
charge of my case and that of the superintendent of the
Hueitien Hospital, had been to the Diocesan House and
demanded fifty ounces of gold for my release and for not
sending the superintendent of the hospital to jail. One of
our colleagues, entertaining them, politely told them that he
had neither money himself nor authority to take any action
but promised to bring the matter up before the Executive
Committee of Three (the committee appointed by me to
carry on the administration in my absence), and said that
they would be given a definite reply two days later at ten
o'clock.

In secret, the members of the Executive Committee held a
lengthy meeting that evening and decided to get assistance
from a number of prominent Christians and non-Christian
friends, including the brigade commander. He became
furious and volunteered to plan a trap at the Diocesan House,
with a number of his soldiers dressed in plain clothes, to

arrest the two men at the appointed hour. Everything was set.

Two days later, when the appointed hour arrived, those two youngsters somehow, instead of coming over to the Diocesan House, went directly to the hospital to see the superintendent. For hours they and the superintendent were in private session in his office while the people at the Diocesan House waited and waited until three o'clock in the afternoon, only to learn that the superintendent had sent them away from his office. Why he did not send them to the pre-arranged trap still remains a puzzle to be solved.

Two days later, Secretary Ming came to our cell and chatted for an hour or so. Suddenly turning to me, he asked, "What is wrong with the superintendent of your Church hospital? I have just learned that there is a possibility of having him arrested. He seems such a nice fellow, courteous and obliging."

"There is nothing wrong with him, but only with the trouble-makers and blackmailers," I emphatically asserted, and told him about the plot of those two youngsters in plain clothes who were trying to blackmail us and get fifty ounces of gold from the Church. "Unfortunately," I continued, "the Church is too poor; furthermore, the Church is not the place for such dirty tricks."

"Oh, is that true? It is outrageous to have such persons blackmail somebody like that! The new regime would not allow that," Secretary Ming commented in a surprised tone. After a while, he asked me, "Can you guarantee the superintendent's behavior?"

"Of course I can. I am willing to guarantee his behavior," I replied. "If he were unreliable and immoral, he would not be an ordained minister in the Church."

"If you guarantee him and he is proved guilty, you will be double-punished—double-punished, remember," Secretary Ming declared.

"Surely, I accept your terms—even thrice punished," I answered unhesitatingly.

"Well, well, this is my chance [chance of getting himself some credits or merits for promotion]. I'll see to it tomorrow," he exclaimed.

Two days later, I learned that Secretary Ming went to the Public Safety Bureau the next day and reported the case of blackmail and my willingness to guarantee the superintendent's behavior. In addition, he and two investigators from that Bureau drove to the hospital seeking to have an interview with another accuser of mine, Yu Ching Chao, who gave his address as the Hueitien Hospital. They talked with the superintendent. Reading over the hospital list of personnel, they could not find a person named Yu Ching Chao among the servants, the staff, the teachers, the students, the nurses, or the doctors. It was a fake! At the same time, the superintendent told them about the blackmail attempt and gave them the names of those two youngsters in plain clothes. They in turn expressed their firm and positive determination to investigate the matter thoroughly and trace the youngsters until they found them, but it all ended in nothing. A few days later I was even warned by Secretary Ming not to talk about it any more. With the exposure of the intrigue to this young and idealistic candidate for the Party membership, and my guarantee for the superintendent's behavior, I suppose the Communist authorities thought it better to delay the arrest of the superintendent and let him manage the hospital for the time being.

Another story of attempted blackmail by one member of

the Youth Corps on another member may be told here. (The Youth Corps is a Communist organization, the members of which are candidates for membership in the Communist Party. The members of the Youth Corps must go through a process of indoctrination and earn a certain amount of merits by working for the Party to qualify themselves for Party membership.) In our cell there was one young rich merchant, named Li Tai, a native of Yunnan. On one of his plane trips to Chungking, the wartime capital, before the change of government, he met Mr. T. Sheng, the real Number One Nationalist agent in Southwest China. In order to get some protection for his business in cotton yarn, Li Tai attempted hard to cultivate friendship with Sheng, but he succeeded only slightly. Because of this little connection with Sheng, Li Tai was arrested by a Communist agent, Mr. Chiang, as his nineteenth victim offered to the new regime. Mr. Chiang was a member of the Youth Corps and was working hard to express his loyalty and to get his Party card.

A few days after Li Tai was arrested and sent to our cell, Chiang, acting very polite and friendly, went to Li Tai's home and consoled his beautiful and charming, although nervous and worrying wife, Mrs. Li, with lots of sweet words and tempting promises of getting Li Tai out of jail. Naturally, Mrs. Li, being uneducated and helpless but faithful to her husband, would welcome any assistance she could get, without caring who gave it. She provided Chiang every convenience and let him use Li Tai's jeep in order to help him get around more quickly.

A week later, Chiang came back and reported to Mrs. Li that he had finally made a good arrangement with the authorities. At first they had demanded 200 ounces of gold for

Li Tai's release and only after he had argued and argued, and testified that Mrs. Li did not have that much money, did they finally come to an agreement for 100 ounces of gold. As soon as Mrs. Li had the gold or money ready, he would get her husband out. Before he went away, he told Mrs. Li repeatedly, "I have done my part. Now it is your turn. Go immediately and get the money ready."

Mrs. Li was very grateful to Chiang and began to unearth all the buried treasure of the family, including her own diamond ring and gold bracelets and gold bars. According to her estimate, they amounted to about half the amount. For the rest, she had to collect from their tenants and borrow from friends and relatives. So she went around requesting and begging her tenants, friends, and relatives for any amount they could provide. It was not easy to raise so much money at a time like that, when business was at a standstill and many shops were closed. For five days in succession, happy and undiscouraged, she trod through the small alleys and big streets of Kunming, practically begging all those whom she knew for enough to make up the other half. She did not let Li Tai know anything about it, intending to do it herself to surprise him.

In the meantime, because of his youth and money, Li Tai was getting along very well in prison with the imprisoned Communist agents. He was granted membership in the Youth Corps and assigned a number of prisoners to observe and check, on whom a weekly report was to be made. He was happy and contented. But one afternoon the housemaid, instead of his wife, came to the outside of the prison with his food. Through the bars of the high window he was able to see the maid and inquired why his wife had not come for the past three days. Was she sick or had she been disloyal?

He insisted that the maid tell him the truth. Straightforwardly the maid said, "A few days ago, Mr. Chiang came
to the house the second time. He and the mistress talked
very quietly in the living room for about an hour. Just for
curiosity, I hid myself in the kitchen and heard that he had
been around for a number of days in your jeep and that,
finally, he had secured a promise of your release for 100
ounces of gold. Since then the mistress has been busy from
morning to night going from tenant to tenant, friend to
friend, and relative to relative, trying hard to raise the fund.
You should be happy and grateful to her."

Having heard this, he jumped down from the high window, furiously shouting aloud to himself, "Lao Chiang, you
are crazy to have arrested me as your nineteenth victim. Now
you want to cheat my wife for my jeep and gold! I am a
worker of the Public Safety Bureau, too, and am not afraid
of you any more. I'll report this and see what the Public
Safety Bureau will do!" From his emotional outburst, we
realized that Li Tai had been appointed as another agent in
our cell to watch and spy on our daily conversation and
behavior. He then ran out and reported the case to the authorities of the Self-Salvation Association. (The Self-Salvation Association is an organization in jail organized by the
Communist agents for the purpose of indoctrinating and
spying on the prisoners.) It was in turn reported to the authorities of the Public Safety Bureau. According to the information Li Tai got later, the case was investigated and the
jeep turned over to the Public Safety Bureau as Li Tai's
offering, but the ransom was stopped. The case was thus
closed and nothing more was heard about it.

9

Communist Care vs. God's Love and Miracle

ONE GENERAL privilege in the district jail was that daily food—two meals a day—was allowed to be sent in by the families of the prisoners. For the first two weeks, ninety per cent had their meals brought in daily, but, as time went on, things became worse and changed rapidly from day to day. After four weeks, the number was almost reversed and later practically all prisoners were dependent upon the food provided by the prison authorities.

The majority of the prisoners were salaried workers and when they were put into prison, their salaries were stopped by the unions. Incomes were cut while expenses increased. Business was at a standstill; at least one-third of the shops were closed. Borrowing from friends and relatives became daily harder and harder. The only thing left for their wives to do was to sell what they had, and for days and weeks sidewalks on the Kunming streets were full of junk, clothes, furniture, and what not for sale.

Families were broken up. Children, soon finding that their parents could not send them to school, became newsboys on the streets but there was no sale. Finally, they had to join the Communist army, the only way of survival. Some wives packed up and went back to their parents' homes, and others were forced by circumstances to live with other men of some means.

It was tragic and heartbreaking to listen to the stories related, with those prisoners crying and weeping all day long. The sad, suffering beyond human limit, refused to accept consolation; the despondent, seeing no hope ahead, constantly attempted to commit suicide. What the Communists said was true: "This is the great epoch when heaven and earth are turned upside down!"

To keep the prisoners alive, two meals a day were provided, and at each meal, every prisoner was limited to two bowls of rice. Also, for every group of twelve persons, they gave a pot of vegetables or soup, which we prisoners nicknamed "glass soup." It consisted of a few leaves of Chinese cabbage cooked in water and once every fortnight—the first and fifteenth of each month—they put three or four pieces of pork into the soup, cut so thin that they were transparent; thus the name "glass soup." It tasted so flat that we prisoners had to put in some of the salt sent to us by our families. Quantitatively, two bowls of rice were enough for the aged but far below the needs of the younger "hungry ghosts." Consequently, the younger fellows just robbed others of their share by eating much faster and helped themselves to more in defiance of orders. At mealtimes quarrels and fights were common occurrences.

The worst of all was that not only was the rice full of tares, sand, and sometimes even pebbles, but was also more or less fermented. For the first week, eating was a drudgery to us. We had to take so much time to pick out the tares and small stones with our chopsticks, and then force it into our stomachs. It smelled and tasted so strange that it usually reversed its course—instead of getting the rice into our bellies we could not help vomiting. Pretty soon, though, on the verge of starvation, we swallowed the rice in the then so-called "revolutionary style," as hungry beasts, paying no attention to either tares or stones, smell or taste. Eating the worst food, living in the worst quarters, and doing the dirtiest work are parts of Communist indoctrination and the "revolutionary style of living." A pound of bread or a few pieces of home-made biscuit constituted a luxury. If you were willing to share them with others, you were admired and respected for becoming progressive. On the other hand, if you tried to keep them for your own consumption, you were envied, and criticized at group meetings for living a feudalistic life alone. Indeed, hunger and culture could not go hand in hand!

When we first entered the jail, we were all surprised to find that a whole room downstairs in the east wing opposite us was given over to a clinic. There was a large and striking signboard hanging down by the side of its door, "The Free Clinic of District Jail of Kunming," with a schedule of daily office hours, except Sundays. I often commented, "This must be a model jail, with provisions of medical care for the sick —something we don't find very often even outside in society. We don't have to worry. If we ever get sick, we may go there and see the doctor."

Some time in the second week, I got a small cut on my left hand from washing the floor. I remembered the free clinic and dashed there to get treatment. After I knocked on the door and opened it, I found a fellow, not more than twenty-five years of age, lying down on a wooden bed. Nearby there were only a rough square table, on which were a few bottles covered with dust, and two wooden benches. So, in my mind, I began to doubt his qualifications as a physician and thought he might be only an attendant. However, I treated him as if he were a physician and courteously asked, "Doc, may I have a bit of mercurochrome? Look, I have a cut here."

Slowly he got up. Raising his eyebrows and looking straight at me, he surprisingly asked, "Mercurochrome? What is that?"

I thought that he did not know the name of the medicine in English and immediately changed it into Chinese, "Hung Yo Shui [red medical water]."

"Oh, 220! Sorry, we don't have it here, but you may have some Tien Chiu [iodine] for your cut. You know, Tien Chiu is much better than 220," said the doctor. "Come over to the window," he continued, "and let me give you some." I proceeded to the window and the doctor, taking one of the bottles of medicine and turning it upside down, slowly pulled out the cork, which was conveniently used to take the place of a stick and cotton, then put a thin layer of iodine on my cut. I thought it was all over, so thanked him and started for the door.

"Hey, fellow, don't go yet. Pay the bill," shouted the doctor.

"Do I have to pay for those few drops of iodine? Don't you say 'free' on your signboard, sir?" I politely replied.

Then the doctor abruptly said, "Free to see you; free to examine you, but not free medicine, which costs us money, too. The bill is five dollars."

The atmosphere was not friendly so without any more argument, I said that I would pay him later and left.

For the sake of curiosity, I began to inquire as to what kind of doctor he was. Pretty soon I found that he was only a quack who had been an attendant at a local hospital for only six months. He was known to the old-timers as the "foursome doctor," having and using only four kinds of medicine—quinine, aspirin, castor oil, and iodine—to treat all his cases. On account of his inadequate training and high charges, his business was gradually fading away while such diseases as colds, headache, malaria, sore throat, tonsilitis, diarrhea, dysentery, indigestion, constipation, and appendicitis were getting worse and worse as time went on in this congested jail, with 407 prisoners from the rank and file of society. No medical care could be easily secured from without and practically all the prisoners were left to themselves and their fate. Anyone in the prison who had a little medical knowledge and would be willing to help others could climb in popularity in no time.

After much thinking and praying, I finally decided to do something along this line, very cautiously. I easily got permission from Secretary Ming (the real boss then) to have all my drugs for family use and some from our Church hospital brought in. In a matter of hours, I became a "quack" physician and nurse myself and was busy with the sick every afternoon. Before I was released, bottles of sulfadiazine, sulfaquanidine, soda mint, cascara, quinine, plasmaquinine, DDT, APC tablets, iodine, mercurochrome, castor oil, aspirin, and multi-vitamin tablets were given out. With my little medical common sense, free treatment, medicine and nursing care, I had no rivalry and was the object of praise and gratitude, respect and love.

Treating any case, large or small, serious or slight, I conscientiously made a habit of saying a short prayer in silence, with my head bowed and eyes closed, as a means of getting some confidence myself and of bearing a little witness to Christ. With my patient and tender care of the sick, God worked miracles. The name "Old Bishop" given to me by the members of the Bible classes became my popular title of respect among all my fellow prisoners in jail. With God's help, I did all I could and did not prescribe wrong medicine, and all the sick got well except two acute appendicitis cases which were beyond me. However, my small knowledge of the symptoms of appendicitis, such as headache, stomach-ache, wanting to vomit, and temperature, helped. Instead of treating them myself, I requested, urged, and pleaded with the prison authorities that they be sent to the hospital. Working through the Communist agents, the Self-Salvation Association and the secretary to the warden, we got permission from the Public Safety Bureau to have them sent to our Church hospital in time for immediate operation, and they were both saved!

In dealing with severe cases, I was sometimes more careful than seemed necessary but, thank God, during my almost seventy days' practice as a "quack," I did not make mistakes. For instance, in the second appendicitis case, a number of fellow prisoners (who seemed to have some medical common sense, too), seeing that a very popular patient, who was much loved by us all, suffered terribly with headache and stomach-ache, requested me again and again, and later even demanded that I give him a dose of castor oil or a few tablets of cascara in order to give him some relief. After three hot arguments with them, I still refused to give him any medicine, particularly castor oil, for I suspected that he was suffering with an acute case of appendicitis. I did not know what to do, but prayed while they were still demanding that I give him some medicine to relieve his pain and suffering. Finally I got up and told them, "I am only a quack but you are all quacks, too. You are Kiang's friends; so am I. We all want to save him, not murder him. To use medicine properly is a blessing but to use it wrongly is murder. I have an intuition not to give him any medicine but to request and plead to have him sent to our Church hospital. Although I am not sure it is appendicitis, I strongly suspect that it is. If it is appendicitis, to give him castor oil is to murder him. If you insist on giving him medicine, you do it and I will have nothing to do with it." Only then were their mouths shut.

Simultaneously, by the grace of God, the authorities of the Public Safety Bureau came to the jail for inspection and I told Kiang's friends to present the case to them with a strong plea, which they did. To our great surprise, the request was granted and Kiang was therefore sent to our Hueitien Hospital. In the evening we received a report that it was indeed an acute appendicitis and that Kiang got there just in time for the operation which saved his life. I was very happy about it and it was a great victory scored over all the other well-intentioned quacks. I was considered a quack no more but a good physician, much against my wishes. However, some fellow prisoners correctly put it, "He is a good doctor because his God is with him."

For the first time in my life a miracle of God happened in the second week of our indoctrination which knocked all materialism and Communism cold. A fellow prisoner and cellmate, named Mr. Tung (who was on my left while my Christian companion, Mr. Hsiao, was on my right), was utterly pathetic, with neither relative nor friend in Kunming, and not a penny or anything in the way of personal belongings. According to his story, he had worked in a store in Kweiyang and had had a nice income—more than enough for himself. As he was afraid of the Communists approaching Kweiyang, he had left the store and come over to Kunming three months before, trying to get a job or some help from a friend of his. After his arrival in Kunming, he found that his friend had gone to Hongkong. In order to survive at all, he had to sell all his personal belongings to pay his hotel bills and meals. Because he was poor and unable to tip the servants at the hotel, he was accused of being a spy when he looked at a plane of the Nationalist government flying over the city. So there he was, in jail, helpless and penniless. His condition was so pathetic that I had to let him share my soap, toothpaste, and what not. I tried constantly to comfort him by telling him of the life and teachings of Christ and His sufferings and death. He seemed to have taken them in and in due time we two became rather intimate, so that quite often we opened our hearts to each other.

One day he suddenly became sick, with a high temperature, and suffered with a terrific pain in the back of his neck on the left side. After examining him, I found no symptoms except a small hump of flesh that was hard and "popped up." A slight touch on it would make him shriek as if somebody were stabbing him with a sharp knife. I gave him some sulfadiazine and soda mint tablets, with lots of water, but they did not work.

Knowing he was penniless and helpless, we were all surprised to see a Chinese doctor come in the next day, who prescribed some Chinese medicine which I had to cook and feed him. Before the Chinese doctor went away, he told us that if that medicine did not help within five hours, it would be better to get a Western doctor. Five hours were soon gone but the medicine did no good, and he was suffering more. Without any request on our part, our cell was unlocked in the middle of the night and in came a Western-trained doctor, who examined him carefully and prescribed the same medicine as I had before—sulfadiazine and soda mint tablets. He took them but they were of no effect. For four days and four nights he lay on the floor by my side, constantly groaning and occasionally screaming. It was indeed a serious case and his suffering increased more and more! It was not only serious to him but also to us fellow prisoners, for later on he was constantly shouting so loud that none of us could sleep at night. Although we all became suspicious of him as he had both Chinese and Western doctors come in to see him so easily and punctually—something that had never been done before—we had no chance to talk to him because he was too sick. For the sake of our close friendship, I did my best to nurse him and wished I could do more for him, in spite of our suspicions.

On the third morning of his sickness, in my daily Bible reading I came across the passage, "With God all things are possible." Convinced that this was a revelation, I prayed fervently for Tung's faith in God and my own worthiness as His instrument. For hours I sat quietly in a corner of the cell, reading and praying alternately. Sometimes I doubted whether it could be done; at other times I questioned whether I was a worthy instrument that God might use. When night came, I could not sleep and was restless, just thinking and meditating, with doubts like Thomas' and fears that I might bring shame to God instead of glory, should I ever try. However, I was reminded again and again of the passages on the power of faith and, furthermore, my compassion for Tung's unbearable sufferings and pain grew stronger and stronger. Early in the morning of the fourth day I crawled over to Tung's place and whispered to him, "We human beings have failed to give you relief, but not God. If you have faith in Him whom I have told you so much about, trust in God's power and love, present your sufferings to Him, and rely on His healing, I am sure you will be all right again."

For half an hour he did not respond to what I said, just lay groaning and screaming—and also thinking, I guess. His sufferings were getting worse every minute. In the hour of despondency and distress, Tung was no exception and, like all human beings would do, responded with a great faith in God, crying, "I do! May your God help me!"

I began to rub the humped flesh lightly at first and then hard with my right hand and prayed fervently and continually for almost an hour, with Tung lying by my side, saying, "I believe." After a while, his pains were miraculously relieved, and Tung became quiet and dozed. At noontime the same method was applied. Everyone in our cell was very much amazed at his speedy recovery and, in the evening, all the cell mates voluntarily knelt down and joined me in prayers of thanksgiving to God for his healing. For the first time in four days and four nights, Tung went to sleep, sound sleep. On the fifth morning, to our great amazement and joy, Tung was the first one to get up—earlier than anyone else. Indeed, "With men, it is impossible, but with God, all things are possible!"

After our first meal twelve of us, including another Communist agent from Cell No. 2, held a short service of thanksgiving to God. To those witnessing this great miracle of God, materialism or Communism was swallowed up in victory. O Materialism, or Communism, where is thy sting? The sting of Materialism or Communism is disbelief, and the strength of disbelief is the pride of man! But thanks be to God, who performed the miracle through His son, Jesus Christ, our Lord.

10

Organizing "Study Groups" for Indoctrination

AFTER SEVENTEEN happy and busy days with my
Bible classes and discussion meetings, the members
were told by the Communist agents that they would have to
stay in jail longer if they continued to attend my classes. For
the sake of safety first, they came to me in groups and individually and advised me to discontinue the meetings, with
promises that they would come to me for more studies and
baptism at our church as soon as freedom was restored to
them. At the same time, an order was issued by the Public
Safety Bureau that no other teachings would be permitted in
jail and that the nine standard books on New Democracy
and Communism must be read and digested before anyone
would be released. The books listed on the bulletin board
were:
1. *New Philosophy of Life*
2. *New Democracy*
3. *Thoughts of Mao Tse-tung*
4. *The Directory of Thoughts*
5. *The People's Democratic Dictatorship*
6. *Introduction to Dialectical Materialism*
7. *Chinese Revolution and Chinese Communist Party*
8. *Life of Mao Tse-tung*
9. *The Present Condition of the World*
In addition, two more books were to be read voluntarily,

namely, *Land Reforms,* and *Communism and China.* The former nine books were more or less pamphlets while the latter two were in book form, thick, and difficult for the majority of people to read. They made reading the last two books voluntary purposely, to give the people the impression that they were in the stage of New Democracy, not Communism.

The members of each cell were organized into a "study group" and there were twenty-seven study groups altogether, corresponding to the number of cells. Each study group elected its own chairman and secretary. The election was very cleverly manipulated by the Communist agents who put up my name, together with one who was pro-Communist, for chairmanship of one group, on the sure assumption that I would not take it. It worked out just like that. I was elected but I politely asked to be excused as being too busy with the sick and, of course, the pro-Communist was then elected unanimously. The secretary was elected in the same way.

In order to pull the wool over the eyes of the members, I was asked to serve as their adviser and teacher, but the Communist agents did all the talking and manipulating while all the rest, including myself, did not even see the necessity of a chairman and secretary in such a group. We did not care how the election came out.

Not until we held group-criticism in the study group could we realize the importance of the chairman and secretary, who were invested with so much power that they could let one prisoner off easily or condemn another, according to the will of the Communist authorities above them. Before the first meeting had adjourned, it was announced that each member was requested to contribute fifty cents toward the Book Fund, and that on top of the twenty-seven study groups there would be an organization named the "Self-Salvation Association." Four fellow prisoners (they were all Communist agents) were appointed by the Ministry of Public Safety to take charge of the affairs of the Association, for the purpose of directing studies and reporting weekly the progress made by each individual.

This was the first time we learned of the Ministry of Public Safety, so we were a little surprised. But later, we discovered that the Bureau of Public Safety was only a temporary and city organization, while the Ministry of Public Safety was a permanent and area (Southwest China) organization. Its Minister, Liu, was appointed by, and directly responsible to, the Communist government in Peiping. In his hands was invested the power of life and death of prisoners. Secret Communist agents were planted in all cells (in our cell there were two) to observe and watch closely every prisoner's words and behavior, and the progress made by the prisoners in their Communist studies, thinking, living, and being frank in conversations and confessions. Carefully and fully analyzed weekly reports of individual cases were made by these secret agents to the Self-Salvation Association which, in turn, with their own observations and comments, if any, turned them over to the authorities of the Ministry of Public Safety.

During the first week after the inauguration of the Self-Salvation Association, no schedule of study hours was announced but books were being purchased, and everyone was encouraged to read as many books on New Democracy and Communism as possible. In two days, with the assistance of relatives and friends outside, the members of our cell got hold of twenty-three books, including, of course, a few duplicates in addition to the required list of books. They were the series of books called, "The Study Series," published by the

Study Association, a newly established publishing house in Kunming.

The members of our cell started to work immediately and every literate member was allotted one book to read and report to the study group. There were altogether sixteen different books or, rather, pamphlets, written in the simplest style, well analyzed and organized. They were easy to read, with possibly three exceptions—*Communism and China, Land Reforms,* and *The Directory of Thoughts.* The last is really the handbook of the Communists, in which all systems of thought in free democracies were severely criticized and only the Communist thinking was considered right and true. According to that, every person was required to criticize his sinful past, be purged clean and regenerated. In a week, I finished them all and was requested to give five book reports to the group instead of one.

In general, after the books had been bought and read a little by the majority of my fellow prisoners, it was found that a few technical terms and philosophical phrases were beyond their knowledge and understanding. Troubled by family problems of starvation and their own personal sufferings, both physical and mental, the prisoners as a whole could not concentrate on studies. Consequently, there was no sign of progress at all.

Conscious of their duties and responsibilities, the group chairmen in their first weekly meeting decided to petition the Ministry of Public Safety, through the Self-Salvation Association, for teachers and lecturers to be sent in to help. The petition was speedily granted and two "progressive" students, both juniors of Yunnan University, were sent in, whose duty was to explain and answer all the questions raised by the prisoners in relation to New Democracy and Communism.

Appearing very self-confident and proud before us prisoners, some of whom were formerly their teachers and professors, those progressive students unfortunately faced in the first meeting a number of questions which they could not answer, such as: explanation of "Concept depends upon reality but reality does not depend upon concept," found on the first page of New Democracy. Also, "Why don't monkeys today become men?" "Is it possible that some day men will become monkeys?" They were dumfounded and contempt was manifested in every corner.

The first instruction period was adjourned by the cleverer of the two by saying, "You have a number of professors among you. Please ask them. They will give you better answers and explanations." So they left, very much embarrassed. After that, no more instructors or lecturers ever came into the jail again, except once, when a high-ranking Communist theorist was requested to come in to explain land reforms to an advanced group of prisoners. Consequently, some of the prisoners, including Professor Tso of Wu Hua College, Professor Ma of Yunnan University, and myself, were constantly besieged by fellow prisoners to answer this question and explain that passage. I was very fortunate in that I had already finished sixteen different pamphlets and books on New Democracy and Communism and, therefore, was ready and able to render service along this line to the less educated and the illiterate prisoners.

11

Communist Grace and the Prisoners' Agony

AFTER THE ARRIVAL of the required books, the Self-Salvation Association hinted that the innocent would be released before the Chinese New Year (February 14, 1950), provided they could prove that they had digested the rudiments of New Democracy and Communism by studying the nine required books. At the same time, Communist agents circulated a report that only 18 persons out of the 317 prisoners in our jail could be considered political criminals, with some evidence against them, and that all the rest were either suspected or innocent.

That hint, together with the circulated report about release before New Year, worked like magic. Happiness and hope were manifested on the faces of prisoners in every cell—happiness and the hope of having a family reunion at New Year time, which is the greatest annual festival we Chinese look forward to throughout the year. Subsequently, during those weeks before this great festival, the spirit of studying among the prisoners was very high. In every cell I visited, I found people either digging into the books or discussing seriously the problems raised by their studies. In a few days, to prove that they had read and digested the books, the prisoners began to send book reports, outlines, and book reviews to the office of the Self-Salvation Association. Soon that small office was filled from floor to ceiling with piles of reports.

The Chinese New Year was drawing near, when a few favored prisoners found out and told the others that that "promissory note" was nothing but a false hope held out by the Self-Salvation Association. Hopes of early release quickly died down and the spirit of study faded away. Such a total disappointment! The whole prison was full of grumbling, despondency, and misery. Passive resistance seemed to be the prevalent outcome, and the authorities of the Self-Salvation Association were left helpless. Time tables for daily study hours, discussion periods, and Yanko (folklore) songs and dances were posted on the courtyard wall, but nobody paid any attention to them. Summons to meetings and talks were answered by only a handful of delegates, who went for fear of missing some hopeful announcements. Orders were ignored. For days, restlessness, quarrels, fights, and even attempts at beating and accusing the prison authorities and those of the Self-Salvation Association became common occurrences, one after another.

In order to redeem itself after making a false promise, and to soothe the prisoners' despondency and end the passive reaction, two days' grace was granted (February 15 and 16— the Chinese New Year in 1950) by the Self-Salvation Association, during which time notes might be exchanged with friends and relatives, New Year eatables might be brought in, and brief interviews permitted with family folks outside the main entrance of the jail through the high windows of those three privileged cells on the second floor of the west wing. Those two days' grace for most of us should have brought some peace and consolation to the prisoners and their families but, contrary to all expectations, it brought only

misery, suffering, and agony. The high window in our cell
was the nearest to the main entrance so it became the most
sought-after place for interviews. We, sitting in the cell,
heard story after story of only sufferings and distress!

On the New Year Eve, agonized children, wives, and
relatives were found outside the main entrance of the jail,
murmuring, scolding, quarreling, crying, and weeping!
Enough cries of sorrow and suffering, and groaning in agony,
were heard to break anyone's heart! It seemed that all the
prisoners were thinking of their past family reunions and
New Year gatherings. What a difference! Now, living behind
bars the life of separation, starvation, wrecked nerves, with
families broken up, they all believed that they were the
victims of dirty politics, bloody wars, and brutal human
nature. Justice, freedom, and hope were gone to the winds!
Some were sitting, lonely, in the corners, others looked idly
at the ceilings, still others wrapped themselves with blankets,
weeping and groaning, and a few angry youngsters paced
back and forth, sometimes kicking the floor and at other
times jumping up in the cells, scolding and beating their
chests. That night no word was spoken in our cell and no
conversation was heard. I sat motionless and dumb on the
floor, leaning against the wall, looking at the high window—
thinking, meditating, wondering, questioning, and searching
for the purpose and meaning of all that I was facing and
seeing that night. I could not make myself at that time utter
a word of comfort or a statement of hope to any of my fellow
prisoners. It was indeed a hell on earth!

A few stories will suffice to illustrate the depth of agony
the prisoners were suffering during those two days of grace.
A few hours after the grace had been granted, a railroad man
with a record of twelve years' faithful service, who had been

imprisoned by his subordinates for his "imperialism" (strict administration), received a short note from his beloved wife who was standing outside. In tears he read, "Our children have been selling newspapers these days, but there is no sale. We have sold everything we had. There is no friend or relative from whom we may borrow as they are all poor now. The railroad union has stopped your salary as you are imprisoned. I have done my utmost. I cannot do more. Sorry for the kids. Take care of yourself. Good-bye!" Early next morning one of his boys came to the prison and told him that their mother had committed suicide the night before when they were sound asleep.

There was another prisoner, about thirty years old, who was once an army officer in the Nationalist army and had contributed greatly to the defense of Changsha in the war against Japan in 1938. He married a beautiful girl from Soochow, Kiangsu, and they had a child. After World War II, instead of going down to Shanghai or Soochow, they came to Kunming and settled down. He had a teaching job, as a military officer, at one of the government middle schools. He was lovable to his friends but very strict with his students. Simply because he had once participated in the Nationalist army, he was considered a criminal and was thrown into prison.

During their seven years of wedlock, he and his wife had never had any serious quarrels, and they were considered by his friends as one of the ideal couples. After his imprisonment, his only income from the school was stopped and their little savings were soon exhausted. Under constant pressure of Communist threats and temptation, his wife finally gave him a small piece of paper on the second day of grace, and asked him to sign it, with a note saying, "To relieve you from worries and to save us [mother and child] from starvation, please sign the enclosed paper." It was nothing else than a divorce document! Suddenly his face became pale and he started trembling. For a few minutes he could not say a word, but at last he shook his head and, picking up a pen, said, with a deep sigh, "Take away my freedom and break up my family! Heaven knows what I'll do. Now I have to sign—yes, I'll sign!"

About four o'clock in the morning on New Year Day, the whole jail was suddenly stirred by a great commotion in Cell No. 4 of the main building. Loud SOS calls were heard but no prisoners could give assistance except those in Cell No. 4, because all the cells were still locked and bolted. No sooner were the cells unlocked than we learned that a former foreman of the 53rd Ammunition Factory in Kunming, disgusted with his own miserable life in jail and bewildered by family problems at home, had decided to give up the struggle and attempted to end his life by hanging himself with a strip of his sheet while his cell mates were tired and fast asleep. However, there was not enough space between the iron bars of the high window and the floor and, consequently, his attempt did not work out well. He tried again by leaning forward and bumping his head against the wall, but that did not work, either. Instead of killing himself, he awoke first one and then all the inmates of the cells, and this caused the commotion. He was saved—saved in time, but saved to suffer more!

During those two days' grace, by feeling agony ourselves and seeing the sufferings of others, we all became very conscious of the value of freedom. Some said, "I wish I could be free with my wife and children," and others commented, "I would fly over the prison wall had I two wings on my back." Among 407 prisoners, only 18 were real political criminals, as the Communist agents had said, with some evidence against them. Most of the prisoners were leaders of society and had had every freedom under the Nationalist regime, including freedom of worship, thought, travel, safe and sound sleep without being bothered by uninvited guests knocking on the door at three or four o'clock in the morning, freedom of silence when they did not feel like speaking, and freedom to cry or weep when they felt sad. As is usual, the majority did not appreciate what they had and remained neutral toward the new regime, some even going so far as to welcome the arrival of the "wolves in sheep's clothing." It was only when they had lost their freedom and were in jail, suffering and in agony, that they began to value and appreciate freedom—but it was too late for regrets!

12

Communist Requirements Before Trials

WEEKS BEFORE the Chinese New Year, the Communist forces, in small numbers under Chu Chia Pi and Chuang Tien, had arrived in Kunming, and the Ministry of Public Safety directly responsible to the Communist regime in Peiping had been set up. The former were responsible for the public order and the latter for the disposal of political criminals.

The New Year's let-down in the indoctrination of prisoners was inevitable because, first, the Self-Salvation Association had failed to "cash its promissory note" of releasing the innocent before the Chinese New Year and, second, it was an annual festival, the first one after the inauguration of the Communist regime in China. During the two days' grace which had been granted, the atmosphere had been pathetic and heartbreaking; the prisoners had felt gloomy and despondent because of the indeterminate aspect of their imprisonment. To the prisoners it looked as if it would be life imprisonment, without any hope of getting released. Consequently, what could be called a sit-down strike came into existence unconsciously.

In order to enforce discipline and to speed up indoctrination, the Ministry of Public Safety took a strong hand, first, by putting a pair of leg chains on one of the prisoners, so as to threaten all the rest; second, by announcing through the Self-Salvation Association the requirement of writing two papers: (1) Tzi Pei Hsu (autobiography) and (2) Tan Pei Hsu (confession), in addition to continued study of the nine required books, and, third, by circulating a report or rumor that all the innocent would be freed soon after the arrival of the so-called "Liberation Army" under Chen Kan, the Commander of all the Communist forces in Yunnan. He was scheduled to arrive in Kunming one week after the New Year festival. The first was a reality and the third only another "inoculation of hope," but the second was something which each of us had to fulfill before we were entitled to a trial and then a hope of release. The trick worked!

The majority of the prisoners immediately started "grinding away" at the books again and tried to find out how to write, first, the autobiography and then the confession. It was then I found out that a small number of our fellow prisoners had written the so-called Tzi Pei Hsu (autobiography) when they first came to jail. From them I got the fixed specifications, and wrote mine. This was an autobiography giving your parentage, date and place of birth, childhood, education, profession, changes of profession with dates, places and causes, cause and date of your arrest, and, last, how to redeem yourself—everything in detail. According to the specifications, this paper was rather easy but was considered important, although not so much as the second paper, called Tan Pei Hsu (confession). In the Tan Pei Hsu you were required to start with your autobiography again and, in addition, you had to give the date and place of your spy training, if any, the spy work you had been assigned to do, the spies you knew and their plots, together with the changes in your reflections on your arrest both at the time of arrest and now, and your "self-conclusion."

Self-conclusion was the most important thing, if you were not involved in politics or spying. It consisted of three parts, namely, self-criticism, self-redemption, and self-determination. By self-criticism you were required to criticize what you had thought and done in the past. Self-redemption meant the work you would do in order to redeem yourself. Self-determination showed the result of both your reasoning and your will power, and the ways in which you would carry out your own redemptive work.

The chief object of this paper of confession was to give every accused person the so-called "opportunities" to confess what he had done in the past that was detrimental to the welfare of the people. The accused was never told the charges against him, except those picked out for examples—he was supposed to make confessions himself. Should his confessions correspond with the charges against him, he would be considered as being frank and honest. With this paper of confession in the hands of the Communists, their judgment of the accused would largely depend upon his usefulness to the Communist cause. Those who could be utilized for the furtherance of the cause would be indoctrinated and sent out to work for the Party, in order to redeem themselves, while those who were of no use whatsoever might be shifted to a labor camp or another prison, or presently disappear.

In order to get prisoners to make frank confessions, strenuous efforts were made by means of speeches, personal talks, posters, etc., so the people would think and believe that the policy of the Communist Party toward political criminals was very lenient. Some popular slogans of propaganda along this line were: "Tell all that you know; tell it in detail;"

"Honest confessors will be forgiven;" "Leniency is granted only to honest confessors." Combined with the spirit of friendliness and courtesy in conversations and at daytime trials, such propagandizing methods were very effective in getting people under confinement and pressure to write out "frank and honest" confessions, with a fond hope that they would be forgiven and released. Some observers often commented, "The Communist leaders certainly know how to apply psychology. Externally they are most courteous and friendly, but internally they are most inhuman and brutal in every sense of the word."

To the Communists, the more secrets you revealed and spy work you did, the more honest you were considered. But the truth was that no matter how frank and honest you were in your confessions, they were never satisfied and always asked for more. Of the confessions of the 317 political criminals in our jail alone, only a handful were considered honest and frank, and these were the confessions of the Communist agents. These were played up as examples and models of public confessions while the rest were asked to rewrite theirs once, twice, three times—some even up to seven or eight times. Every time a prisoner was asked to rewrite his confession, advice or, rather, a warning was given, saying, "Another chance of leniency is given to you. We shall be glad to help you but we cannot help you unless you help yourself. We cannot save you unless you save yourself. To help yourself is to save yourself. To save yourself is to be utterly frank and honest in your confessions."

Another point of interest was that those who were considered frank and honest confessors were soon given a trial or two each, and subsequently released. Those thus released usually, and purposely, made a big show in prison before they went out. Some of them would come (rather, were ordered to come) back to prison in a day or two to visit their fellow prisoners and tell them that the only means of salvation—of getting released—was to be utterly frank and honest in their confessions. Actually, they were real Communists or Associate Communists themselves, imprisoned to spy on the real prisoners and then tried and set free as examples and models of frankness and honesty in making confessions, so their fellow prisoners would follow and be trapped.

Another purpose in requiring the prisoners to write their confessions so many times was to catch discrepancies. After going through so many weeks of mental and physical sufferings, the accused became weak and unbalanced physically and mentally, and dull and sluggish. Made-up stories could not go through the fire of comparison and examination. Discrepancies stood out as clear sky through the glass of comparison. If any discrepancies were found in the confession of any person, he would again be investigated, questioned, and cross-examined. I found again and again that the more the accused tried to explain those discrepancies, the deeper he got himself into trouble. It is very true even with the Communists that only "truth . . . will endure forever," and "Truth shall make you free"—free of all unnecessary complications and troubles.

Following this Christian principle, I was very fortunate in that I wrote my Tzi Pei Hsu (autobiography) only once and my Tan Pei Hsu (confession) four times. Instead of rewriting the whole confession, I was required only to write a supplement each time, such as supplement in detail on my education, on my connections with other denominations, particularly with the Roman Catholic Church in China, on my relationship with the American Episcopal Church, and the Church of England. Grace of God was then clearly manifested that our diocese of Yunkwei (though a Warphan —orphan of war—or work organized during the time of World War II and which had been refused official connections with the mother church in either the States or England) remained the ONLY indigenous diocese of the Holy Catholic Church in China. Although many a time we felt lonely and deserted in that far distant corner of Southwest China, I did reap at least once, thank God, the sweet fruits of being "indigenous" and "autonomous."

13

Communist Judges and Trials

FOR THE FIRST seven weeks there was no trial held at all. This was called the study and self-examination period during which, on one hand, cases were allotted to Communist judges for investigation. Most of these were high school and college students under the direction and supervision of a few Communist judges, formerly trained in Yenan, Shensi (the Red capital between 1934 and 1949) or elsewhere, and were directly responsible to the Ministry of Public Safety of the Communist regime. On the other hand, the prisoners were busy studying the Communist books, writing book reports and reviews, discussing them in study groups, and writing their autobiographies and confessions again and again. Autobiographies and confessions were collected and studied carefully by the authorities of the Self-Salvation Association. Comments were written on the back of each paper and they were then turned over to the assigned judge. Only after all this were our long-hoped-for trials started!

Trials were divided into two classes; the unimportant cases were tried in the afternoon while the serious were at night. For the afternoon trials, each judge was assigned a small room outside the inner gate of the jail proper. The judge called his assigned prisoners one by one for trial. Sometimes a teacher was tried by one of his students; at other times, a

86

manager of an institution or bank was tried by one of his subordinates. As a rule, judges were very friendly, cordial, and courteous to the accused, sometimes so much so that their artificiality was not hard to detect.

The judge sat down on one side of a desk in the room and the one being brought to trial was often asked to sit down on the other side. Not infrequently, cigarettes were offered to the accused if he was a smoker. They were so friendly and congenial that once in a while, deceived by their attitude, a prisoner took advantage and argued with the judge, sometimes even scolding him or beating on the desk between them. No matter what the accused might do, the judge always remained calm and friendly externally. However, the truth was—as we later found out—that one prisoner, because of his anger at the judge, was chained with a heavy leg chain and another, for scolding his judge, was shifted to another jail at night for disciplinary purposes.

Every trial was started with the autobiography of the accused, which was to be told clearly and continuously, from the time of his birth to that of his arrest. Any hesitation or reluctance in telling the story was taken as a sign of dishonesty or an attempt to hide something from the judge, then he would be questioned and cross-examined. Occasionally he would also be asked to repeat something in detail or to repeat things over and over. Autobiography was followed by a long list of questions, prepared by the judge beforehand from investigations and comparative studies of his confessions written at different times or of his confessions with the confessions of others, if in any way related. From the afternoon trial, the judge generally hoped to secure not so much more facts about the accused as facts and evidence against other prisoners in other jails who had been associated with the accused. This was called "the horizontal examination" or investigation.

The last item of such trials was the accused's reaction at the time of his arrest and whether he felt the arrest was justified now, after he had been indoctrinated. No charges against the accused would be told during the trial; he was expected to confess them both in his written confessions and orally before the judge. If he made no reference to the accusations against him or his confessions did not correspond with the charges, his trials, no matter how many would be held, would not help any. He was considered as not being frank and would remain in prison or be shifted to some labor camp. As a rule, the afternoon trials led to no conclusion or sentence, either good or bad, and the accused was sent back to prison with a request or, rather, a friendly order to be patient and examine himself more severely on what he had done in the past that was detrimental to the welfare of the people.

The serious cases were tried at night. Every evening was mental torture to all of us prisoners. When the cell doors were locked and bolted at 8:30 p.m., night trials began and almost every inmate became frightened and nervous. No one knew who would be taken out and tried that night. Anyone to be tried at night was always blindfolded and taken by two guards to a room outside the gate of the jail proper, where usually four or five judges were ready to take their turns to question, examine and cross-examine him. For hours and hours the accused had to stand there blindfolded and bombarded with all sorts of questions.

Whenever the Communists felt it necessary to get confessions from the accused for some purpose, they never hesitated to use threats, beatings with clubs or bamboo sticks, or tortures such as the Tien Hua Chi (literally translated, "Electric Telephone Machine"). This was a simple device consisting of a wooden box in which a battery or two was placed. It was electrically charged and installed, with two cords, the ends of which were attached to the prisoner's two hands or other parts of the body. The one who operated the machine turned the handle of a wheel so as to give the tortured person electric shocks, sometimes continuously and sometimes with breaks, according to the orders of the chief judge. Anyone who had gone through that electric torture was always ready to confess whatever they desired. This was the most commonly used torture in our jail.

As a rule, night trials lasted from nine o'clock in the evening until three or four in the morning. In no case was a prisoner coming from the night trial happy, but was always sad, quiet, worried, listless, and not infrequently half-dead.

In the beginning of my eighth week in prison (the end of February 1950), trials began to be held in our jail. I had my first trial the same as everybody else. One afternoon my name was called and I was told to go to the living room of the warden's quarters, which had been turned into courtrooms. I was tried by Judge Chien. When I entered his room Judge Chien, very polite and friendly, even stood up and asked me to take a chair and sit on the side of the desk near the door. As a regular routine, he started the trial by asking me to tell him my autobiography. I told him my story exactly as I had written it, and continuously, without a stop or break to think of what I should say.

At the end of my story, Judge Chien asked me to narrate in detail my education, which he wrote down himself very carefully. It was the same thing I was required to write as a supplement to my written confession. After that, I brought up the charges against me published in the Ping Ming Daily on January 5 and requested, with a strong plea, that the members of the Kunming Christian Fellowship be brought in as my prosecutors, with evidence. Then I also asked him, "Will you please tell me the members of that Fellowship? I have never heard of the organization. We Christians in Kunming have only one organization, called the Association of Christian Organizations in Kunming, to promote fellowship among ourselves and our united Christian service to others. Our motto is: Agree to Differ; Resolve to Love; Unite to Serve."

Hearing my plea, Judge Chien just smiled without saying a word. After a while he said in a friendly tone, "Don't worry about that. We know something and are still investigating. Now, answer my questions, please." He opened his file again and picked up a long list of questions, prepared beforehand, which he asked and I answered one by one. Some of the questions were: "Is your Church democratic or imperialistic?" "How is it organized?" "Are the missionaries imperialistic?" "Some missionaries look to me as if they were not qualified to be Christians. How is it that they have been sent to China as missionaries?" "What is their real motive in coming to China?" "Why don't they give equal treatment to the Chinese workers?" "Do they send reports about China back to their governments?" "Why don't they respect the Chinese Church leaders?" "Why is it that the missionaries in Kunming do not speak well of you?" "Don't you feel that you are obliged to take orders from those who have been giving you funds for your work?" "Is it necessary to have so many denominations with you Protestants?" "Is it possible to organize all the denominations into one Church?" "If so, how?" "Can you get along without foreign missionaries or funds from abroad?" "What do you think of the missionaries of your Church [giving me their names in Chinese]?" "What do you think of the missionaries of other denominations [giving me also seven or eight names in Chinese]?" "Have the Rev. Messrs. Liu and Kong done anything harmful to the people in the past?" "What is your special connection with Dr. Brown?"

Hearing these questions, I was totally surprised and amazed at his knowledge of the churches and their leaders in Kunming, so I said to him, "You really know much more about the churches in Kunming than I do, for instance, some names of the foreign missionaries of other denominations that I don't know at all."

Putting down his pencil on the desk, he looked at me and smiled, saying, "That is no wonder; you haven't been here very long."

To my answers his reactions varied a great deal—surprise, agreement, disapproval. For instance, my explanation of the questions, "Why don't foreign missionaries speak well of you?" and "Don't you feel that you are obliged to take orders from those who have been giving funds for your work?" stirred up a little surprise in him as I answered, "It is not surprising that they don't like me. This is partly due to jealousy; partly due to their loyalty to the society which has sent them over instead of to the Church in China; partly due to their desire to make the hospital an independent institution instead of a means of evangelism; partly due to our differences in opinion, policy, and method, and largely due to the fact that I have been fighting hard for equality between foreign missionaries and native workers. To those who have been giving us funds for the work, I am grateful indeed, but only in behalf of the diocese. Personally, I don't feel I am obliged in any way at all, because they give funds for the work, not for myself. On the other hand, the donors should feel grateful to God for their blessings and the privilege of being able to give, and to us who are actually doing the work, in a sense, for them. What is right, I do; what is wrong, I don't do. I pay no attention to them no matter what they say behind my back or what pressure they try to bring on me. I am *not* their 'yes man' or 'running dog.' "

Of my answer to his question, "Is it possible to organize all the denominations into one Church?" he showed an enormous amount of disapproval and impatience when I explained, "What we need is unity, not uniformity. Variety is the spice of life! We human beings are brought up with different backgrounds, different environments, different degrees of education, and so on. We cannot put all beings into one mold, expecting them to be in the same shape and take the same form. This is what we have been trying to do by means of the Association of Christian Organizations in Kunming. Whatever we do among Christians of various denominations is guided by the principle: "agree to differ, resolve to love, and unite to serve."

Before I finished answering all his questions, two hours and a half had passed. Judge Chien, after looking at his watch, said, "I have one more trial to conduct this afternoon. We know your case pretty well. Don't worry! The new regime is always just. Be patient! We need some more investigation; in a few days you will be all right!" He got up from his chair and escorted me to the door. So came the end of my first trial.

His knowledge of the churches was a great surprise to me. For the sake of curiosity, I tried to find out who Judge Chien was. Soon, from two fellow prisoners, both natives of Kunming, I learned that he was a native of Hopeh and had been in Kunming for years working as an underground agent of the Communist Party. His assignment was to investigate and record the tendencies, utterances, and everything possible about the leading educators, missionaries, and Church workers in Kunming. Indeed, he knew his stuff—well done, a faithful worker of his party!

Based on these few words of comfort from Judge Chien, "In a few days you will be all right," my spirit was high again, expecting my restoration to freedom at any moment. Unconsciously I was smiling and singing! But hours and days went by and nothing happened. Recalling the repeated brutal treatment of some of our fellow prisoners—beatings, leg chains, and tortures at night, while friendliness and courtesy were shown in the daytime—and remembering those sweet "promissory notes" to us all which were never "cashed," I began to feel skeptical. Soon I became down-hearted again; my hopes of release faded away! Truly indeed, the real Communists "come to you in sheep's clothing, but inwardly they are ravening wolves." Beware!

14

My Accusers and Second Trial

DURING the week following my first trial, I was happy, expecting my early release, as Judge Chien had said, "In a few days you will be all right," but I had forgotten another remark of his, "We need some more investigation." Almost every evening I was visited in my cell by someone from the Self-Salvation Association or some Communist agent for a chat, for a copy of the New Testament, for some more information about the Church and its teachings, for an explanation of this passage or that in the Bible (as he liked to know about Christianity), or for some knowledge about such and such denomination or such and such leaders in the country.

I talked, and talked freely, without the least hesitation, for I felt that "real gold would not be afraid of fire." But once in a while I became quite conscious that it was an investigation rather than a conversation. In such conversations, I was hoping on one hand, by giving them facts about the churches, to help them correct their misconception of the work of the Church and live up to the proclaimed policy of the new regime—freedom of religion—and on the other, I was anxious to find out from them more definite information about my accusers—the members of the so-called Kunming Chris-

tian Fellowship. That is why I purposely went into detail
with my answers and explanations.

One evening, in the midst of our conversation, Mr. Chen
became daring and talkative, and, realizing that the person
with whom I was talking was a Communist agent, asked,
"Isn't it detrimental to the cause of revolution that so many
innocent people have been arrested, including a bishop whom
we have come to know so well? What do you mean by
'freedom of religion?'"

Quickly trying to defend the new regime, but unconsciously giving out some information which I had been very
anxious to get, he replied, "As to the Old Bishop, we cannot
blame the new government. His trouble lies in the Hueitien
Hospital. They think that he is a dictator and imperialist,
with his episcopal power. [According to our canon law, the
bishop does have the veto power but I never used it during
my active episcopate.] A lot of people connected with the
hospital want to have it separate and independent, but he is
bitterly opposing it. Well, at any rate, the new regime will
not tolerate such hospitals trying to buy the hearts of the
people under the cloak of medical service. We know his
troubles. That group, including a few foreign missionaries
and six or seven Chinese, forms the Kunming Christian
Fellowship. They are his accusers—that bunch of rotten and
cowardly fellows! I have heard that when they were approached, no one claimed to have any connection with the
Kunming Christian Fellowship. Perhaps Old Bishop is correct in saying that the Kunming Christian Fellowship is a
non-existent organization. The new regime is certainly treating the old bishop nicely. He was arrested, *not for his faith,*
but for something else. We certainly give freedom of religion
in the Common Principles." At the conclusion of his answer

to Mr. Chen's question, he suddenly became aware himself that he had said something which he should not have said, so he repeatedly warned us to keep it confidential among ourselves.

That talk was certainly revealing to me. For the first time I learned that the new regime would not tolerate such hospitals (all the Church hospitals) trying to buy the hearts of the people under the cloak of medical service. With some further information I received later, I came to the conclusion while I was still in prison that the new regime would take over all Church hospitals sooner or later. Words of advice were passed to the superintendent of our Hueitien Hospital to let our missionaries evacuate, and to give up the hospital to the government at the earliest possible opportunity. My advice was ignored and at that time I was considered foolish and crazy. However, not many months later, when the Communist regime imposed heavy taxes on Church hospitals throughout the country, I stood firm in my conviction and not a cent was paid for the irrational business tax ($20,000 U.S.) on the Hueitien Hospital in Kunming for 1949, while many other church authorities did comply and paid heavy taxes not only for 1949 but also for 1950. In the end, the fate of Church hospitals turned out the same, although my "craziness" did save the diocese twenty thousand dollars.

Secondly, his comment that I was arrested not for faith but for something else, and that the new regime was treating me nicely, also reminded me of all the stories I had heard of how the prominent Christians and Church leaders in Northeast and North China were imprisoned and killed. Trumped-up charges, either political or economic, were the causes of their death, for the Communists did not like to see any martyrs of faith. The new regime was treating me nicely—

in comparison with those whom they had decided to liquidate soon—but by being nice, I was only classified with those prisoners whom they wanted to utilize first before they were to be liquidated. Without exception, we all had to go through the same mental "goose-stepping" and physical hardships so that we might be brought in line and be used to the glory of their Party-State. This was confirmed later by their tempting offer.

Thirdly, although I was shocked by the implication that a few missionaries had joined with some Chinese in the plot against me, I partly believed it because I had fired a few workers from the hospital for dishonesty and inefficiency. I also recalled immediately what Mrs. Violet Phinney had said just before she took off for Hongkong, "I wish I had been working in Kunming and could have straightened out those false stories circulated about you by a missionary in the foreign community." Furthermore, I recalled also the words sent in by a clergyman of our Cathedral that Mr. Evans wanted to become the superintendent of our Church hospital again. No matter what the motive of my adversaries might be—revenge, jealousy, or what not—I could not help laughing, and said to myself, "The hospital is a piece of hot coal. Only the Communists have the proper folk to take it and utilize it. Anyone else who tries to take it into his hands is bound to be burned. How ignorant they are and unable to tell the signs of the times, yet claim themselves as 'old hands of China' for having been in China all their lives!" If they ever did plot against me, I had no malice toward them but felt grateful.

Being imprisoned and cut off from all my beloved ones, friends, colleagues and the brethren of the Church, I experienced the imperative value of faith in God. I widened my knowledge of human sufferings, and only through sufferings I learned the meaning of that perplexing problem of life. It is very true that "Whom the Lord loveth, He chasteneth." And "We know that all things work together for good to them that love God." At the same time, knowing the Communist tactics—creating hatred and clashes among others while they stand on the side line watching and reaping profits—I still suspect that it was all a Communist-prepared plot for bringing me in line to work for them for a time. This remains still a puzzle only the future may solve. However, my curiosity was aroused and, if there was any possibility at all, I wanted to get more information.

Four nights later Cheng, who was an easy-going youngster ordinarily but who worked hard for the Party as a probational member of the Youth Corps, came to me and requested me to write a book report for him and answer a number of questions on New Democracy and Communism, as he had to hand this in within a few days. I consented and, in return, I presented him with a request for more information concerning the members of the Kunming Christian Fellowship who had plotted against me. Cheng straightforwardly accepted the bargain and promised to do his best. Days and nights went by; no report came. In the meantime, however, Cheng was seen mingling more and more with the Communists, presumably to obtain the promised information.

One evening, after a lengthy group examination meeting, we were all tired and exhausted and were resting on the floor. All was quiet! In came Cheng and pulled me to the southwest corner of the cell, saying, "I have something for you. This evening, when you were holding your group examination, we held a meeting in which your case was brought up and I was asked to give them my observations on your behavior. You know what I said about you—all good! They said, 'The charges against you came from two sources; one named Yu Ching Kuo, with his address at the Hueitien Hospital, and the other, the Kunming Christian Fellowship, with no address at all. According to the results of our investigation, no such person has been found at the hospital and no one has come out to identify himself with the Kunming Christian Fellowship. However, one of the Communist agents said that he knew who they were—three or four foreigners and four or five Chinese. One foreigner is an old man with white hair, the second a short fellow, and the third has his residence in Shih Chiao P'u, where they meet once in a while. Most probably that foreigner has a wireless station at his house. The Chinese are former employees of the Hueitien Hospital whom you have discharged.

"By the way, what have you done to Principal Yeh? He is in that group, too. He also said that they had been approached one by one but they had all flatly denied any knowledge of the Fellowship. Don't worry, we have sent in good reports about you. Everything will be worked out all right for you!"

While I was listening, I felt rather sad, not because I was in jail suffering, but because we Christians fail to live up to what we believe. That is why our Lord warned us, "Not everyone that saith unto me, 'Lord, Lord,' shall enter into the Kingdom of Heaven, but he that doeth the will of My Father which is in heaven." It is at least partly due, I believe, to the failure of us Christians to build the Kingdom of God on earth that the Communists have sprung up, trying to build their Classless Utopia. After all, we are all sinners before God!

On March 7, 1950, my name was called for my second

trial. I was led to the same courtroom by a guard, but this time the judge was a Mr. Jen, a native of Shantung Province and a big shot in the Communist hierarchy. He was Deputy Minister of the Ministry of Public Safety in Southwest China. On entering the room, I found him sitting there, a man about forty years old, very dignified and slow in speech, with a young secretary sitting at the next desk and recording every word I said. Once in a while he interrupted and told me to speak slowly. As a regular routine, the trial started with my relating my whole autobiography, followed by another series of questions prepared beforehand by Judge Jen, who asked, "Do you still believe in God?"

"Yes, I still believe in God and my faith has been strengthened much more by my experiences in jail," I answered.

"How do you know there is a God?" he asked again, in a slow and deep tone.

"I know there is God from my own experiences," I answered quickly, "I don't need any more argument for His existence. Since I came to this jail, I have experienced more and more. He has answered all my prayers, though most of the answers have not been in the ways I expected and wanted, but in much better ways."

Looking straight at me, he raised his voice and asked, "How is your God protecting you now in jail?"

Without the least hesitation or thought, I replied, "It is not a question of protection but a question of giving me opportunities in jail of learning so many valuable things in life, such as the purpose of life, the meaning of suffering, the teachings of New Democracy and Communism, the pleasure and privilege of service to others, which I would have no chance of learning and experiencing myself if not in jail." At the conclusion of this answer, he stopped and kept quiet for a few long seconds, while in my mind I was offering my thanks to God because that question was not put to me during my first few days of imprisonment when I was really suffering terribly, both mentally and physically, my faith was shaking, and I was wondering whether God had forsaken me. Since then, I had had plenty of time for meditation and prayer and long before my second trial, I had found my answer to *that* question.

Turning to another subject, he asked me, "How much property have you, such as land, houses?"

Easily I replied, "I am a 'propertyless of the propertyless,' owning 'not a single roof tile above, or an inch of land below.'" As he was very much surprised at my answer, I continued, "Please investigate." Later, after my release, I learned that he did send two fellows over to the Diocesan House and asked our janitor and a few others, including Mr. Chu, who answered, "Our Bishop is a poor fellow. He has three pairs of trousers and, to meet my need, he has given me one pair." We had formerly regretted the loss of four homes since 1939, as well as our personal belongings, but it turned out to be a great blessing for me, for now I could not be counted as a landlord or property owner. Blessed were the poor, for they had no worries or worldly complications!

He continued asking me questions: "Are missionaries imperialistic?" "What kind of information do they send home?" "Are missionaries good and friendly to you?" "What have you learned in your study group?" "Do you believe in the Common Principles?" "What are the similarities and differences between your religion and New Democracy?" "What are your present reflections?" When he had finished the questions on his prepared list, hours had gone by, and he showed fatigue himself. So came the end of my second trial, without any conclusion, either good or bad.

15

Intensive Indoctrination

AFTER TRIALS had been inaugurated, and all the required nine books had been read and book reports written and sent in by the prisoners, the "study groups" were ordered to conduct "critical discussions" on the required books, one by one, according to the elaborate questionnaires and syllabuses prepared by the Self-Salvation Association. The chief object of these critical discussions was to draw out personal and inner reactions and reflections of each prisoner on their thoughts regarding New Democracy and Communism.

In the discussions, every member was exhorted (in fact, required) to speak freely. Every empty space inside the jail was hung with posters, "Tell all what you know; tell them in detail." "Those who speak out freely commit no wrong." "We want to help, but cannot help the silent." "Reticence is the sign of disagreement and disapproval."

At first, lots of us thought we were actually given freedom of speech and spoke very freely at discussion meetings. Criticisms after criticisms of New Democracy and Communism were piled up; questions after questions were fired at group leaders who were unable, though being pro-Communists or Associate Communists, to give satisfactory answers to such questions as:

"Communism believes in Classless Utopia. How do you build it on earth—by killing all the class-minded people? They teach, 'We must do according to what we know or believe.' They are killing the people of the old classes, yet building new classes among themselves, such as Communists of small kitchen, Communists of medium kitchen, and Communists of big kitchen. [Note: The Communists in China divide themselves into various ranks: the ordinary members eat the poor food cooked in the government big kitchens; at certain ranks, the members are entitled to get married and to better food cooked in medium kitchens; and at higher ranks, they are entitled to have personal and private kitchens called 'classes of small kitchens.']

"Why don't they do according to what they know and teach? Are they building a Classless Utopia, or another class utopia?"

"We are taught that human beings are descendents of monkeys. Why monkeys? Have the Communists found the missing link in evolution? Why don't monkeys to-day become human beings any more? Is it possible that we human beings may become monkeys?"

"Communism is materialism; both Communism and Materialism oppose idealism and spiritualism. Why, then, do they still talk about ghosts? One of their slogans is, 'The former government changed men into ghosts, while the new regime wants to change ghosts into men.' You are an ardent believer in Communism, teaching and preaching Communism. Why were you afraid of ghosts the other night when you went out into the dark and yelled for help? Ghosts are non-existent. Why do you still believe in them? Is that belief of yours subjectivism, idealism, or spiritualism?"

All these questions and many others put our group leader

on the spot and he did not know what to say. The secretary was busy recording all the questions and comments in his minutes, and the Communist agents were putting down on record cards the names of those who raised questions and spoke so bitterly against New Democracy and Communism, with remarks, "Thoughts reactionary" or "Behavior stubborn." Pretty soon we discovered that we had neither freedom of speech nor freedom of silence, and that New Democracy and Communism were dogmatic and intolerant of any criticisms except those against other systems of thought. After this lesson we all learned how to speak "progressively." As a result, it created in most of the prisoners a double or divided personality—speak in one way and believe in another.

Among the required books of study, the book called *The Directory of Thought* was considered the most important, in which various systems of thought were presented in twisted form and severely criticized. Only the thoughts of New Democracy and Communism were scientifically and objectively true, and served as the only standard by which every prisoner was expected to get his former distorted and sinful thoughts straightened out or purged.

The Directory of Thoughts was ordered to be studied very carefully by every study group, as well as by every individual. It was divided into four sections, with a syllabus of more than one hundred questions for discussions in groups and for answers in writing by individuals. The fourth section of the book was on "Self Examination" and "How to Write Self-Examination." After the completion of the study of this book, including the specifications on "How to Write Self-Examination," every prisoner was required to write his own Tzi Sheng Hsu (self-examination), the third and last paper.

According to its specifications, he started with his autobiography again. This was to be followed by the story of his father's profession, together with the thoughts of his class of people, and criticisms. By giving facts about a parent's life and profession, it usually resulted in his arrest, if he were alive. Then *he* was required to state clearly his own inherited ideas, beliefs, sentiments and viewpoints, including his changes in life due to environments and changes of professions and, subsequently, any changes in his ideas, beliefs, sentiments, and viewpoints, carefully analyzed and thoroughly criticized. In doing this, he had to be absolutely frank and honest in finding the roots and causes of his past distorted conceptions and sinful thoughts, manifested in his daily life—some imperialistic, others feudalistic, and still others bourgeois. It ended always with the ways and means, and his steel-firm resolution to redeem himself. This was what the Communists called Chi Ao Tu Cheng, the "relentless war against self," or self-criticism—an important step in thought transformation.

Self-criticism without group criticism was valueless. In no case would your "Self-examination" paper be accepted by the authorities of the Self-Salvation Association without submitting it first to the study group for thorough examination.

After you had completed your "self-examination" paper, you presented it to the group leader, whereupon he would call a "group-examination" meeting in which your paper was read, section by section, and then studied and mercilessly criticized by the members of the study group. On one hand, you were scrutinized according to the analyzed criticisms found in *The Directory of Thoughts* and in comparison with your own daily life; on the other, you were expected to accept their "friendly and loving" criticisms gracefully and gratefully, with a solemn promise that you would redeem yourself to the best of your ability.

Personal dignity, individualism, old nationalism, heroism, neutrality, idealism, and any sign of "bourgeoism" or feudalism were mercilessly condemned and were to be replaced by collectivism, new patriotism, new nationalism, new internationalism, and scientific materialism, which constituted the basis of New Democracy and Communism.

Such examination meetings were conducted with a great deal of religious fervor, more or less similar to some revival meetings. Many a time the person being criticized got up cursing his forefathers, environments, and the old educational system for his past distorted and sinful ideas, and beat his breast to show his absolute determination and efforts to save himself and then others. This was the Communist idea of repentance and a sign of regeneration!

Such group criticism meetings were long and tedious and lasted, usually, for hours and hours. By the time you got through such a meeting you would, if you were a conscientious person at all, suffer terribly mentally and groan for days. Silence and distress were the outcome. In the eyes of the Communist, we were all sinners because of our heredity, environment, and education and this mental ordeal or torture was the means by which the group would help the individual become "regenerate and make progress."

To meet such mental tortures, however, a trick was discovered by most prisoners themselves, as two cell mates confidentially commented: "We had better face the reality of life and be clever ourselves. The best way of getting through the mental ordeal is to be actors, whether we like it or not, and do our best in acting out the Communist play of repentance and regeneration on the stage." Some were good actors while others were not and, in almost every case, by living with them day in and day out, it was not hard to find out that their conversion to Communism was only in words, not in their hearts. Real converts could not be won by imprisonment, torture, force, or in any way that was against human free will of choice!

When my "self-examination" was brought up for scrutiny, I was very much worried and anxious, for I could not be an actor in the play. However, I got up in time and expressed my willingness and hope that they would criticize me and my paper mercilessly, so as to help me see my faults and redeem myself. Someone suggested that in order to save time, I should read my paper all through once, so I did. After the reading, another cell mate got up and said, by way of criticizing me, "The only trouble I find with him is that he showed no interest in politics, as he has written in his paper of self-examination. Look at the great work he has been doing among us, his fellow prisoners. He has healed many diseases and saved at least two lives. Every afternoon he is ever so busy with the sick and patiently takes care of them. If anyone deserves the title, 'man of service,' it is he!"

Much to my surprise, then came voices in unison saying, "Yes, he is the man of service!"

A third person continued, "We cannot blame him or criticize him. See how much time he spends every day to serve others. He helps us not only with our illnesses but also with our studies. I have never known him to refuse anyone who comes to him for help. Furthermore, he is honest and frank; he is disinterested in politics, as he has confessed in his paper. We should excuse him."

Another roaring voice from the group said, "Yes, we should excuse him."

Immediately I got up and expressed my deep gratitude to them for their appreciation of my medical and other services to them. I further said, "This is the duty of every Christian, 'not to be ministered unto, but to minister,' which, incidentally, corresponds to the idea of New Democracy, 'Service to the people is the object of life.'"

The group leader, or chairman, then started to talk, saying, "We have been with our Old Bishop about two months now and we know him pretty well. I agree with you that I have not found anything wrong with him, either, except his total disinterestedness in politics, which I guess he will change as time goes on. But I would like to call your attention to the fact that real Christians are Communists. So was Jesus a proletarian, having nothing himself but giving all he had to help the people."

Thus came the end of the group criticism of my self-examination paper. Sitting in my corner, I was very happy and grateful to God for helping me pull through this mental torture so easily. However, I could not help disagreeing with our chairman on his statement that real Christians were Communists, and said to myself, "To be saved, Communists must become Christians. Hatred and class struggle can never save. Only by the grace of God and love of Christ you may be saved. There are similarities, yet great differences."

Before such a long and strenuous meeting adjourned, the secretary was constantly busy writing the minutes of the meeting and the criticisms in detail of the person being scrutinized. The group leader also raised a number of points concerning the person for discussion and votes—such points as his understanding of his own thought-problems, his frankness and fluency in answering questions, his politeness, his willingness, his gratitude in accepting others' criticisms, his conviction of wrong and determination to reform and re-generate himself in the days to come. Votes were taken on each point and from the votes, the group leader was to write out his remarks, comments, and conclusion on the back page of the self-examination paper.

When all this was done, the self-examination paper, to-gether with the minutes of the group examination meeting, was to be turned over to the authorities of the Self-Salvation Association. One of the authorities of the Association was assigned to go over the paper carefully, particularly the comments of the group leader. Based upon his personal ob-servations of the person in jail, together with what the person had written about himself and the comments of the group leader, he made a supplementary note here and put down a question there to be answered. Usually the paper was sent to the author, who had to re-write his self-examination paper again and again—from three to six or seven times. When that paper was in final form, it was sent to the authorities of the Ministry of Public Safety, together with the comments made about him by the authorities of the Self-Salvation Association.

Communist indoctrination means not merely "brain wash-ing," or mental purging, but also takes in manual labor as a means of regeneration. This is one of the points often missed by various writers on Chinese Communism. The Commu-nists believed hard labor would slowly wear away the stub-born nature of man, as well as his personal dignity and self-respect. A great deal of stress was put on the doctrine of Unity of Rationalization and Action. To them, action with-out rationalization was blind, while rationalization without action was false. Whatever you believe, in dialectical materialism, must be carried out. As St. James said, "Faith without works is dead."

In jail, various kinds of manual labor were provided and assigned. The higher you were in society, the worse you were assigned to do in order to help you—as they often told us—become regenerate and make progress. When we were at the climax of our mental indoctrination, the authorities of the Self-Salvation Association devised all kinds of competition in manual labor for the prisoners to participate in, either in groups or as individuals. On one day a competition was held in cleaning the gutters; on another in sweeping the courtyard or washing the cell floors; on the third, in cleansing commodes or water closets; on the fourth, in carrying water, killing bedbugs, or transporting bricks from one place to another, and so on. Competitions were carried on vigorously every day. Heroes of this, heroes of that, were voted, announced, and praised as models for others to follow. By one's willingness and hearty participation, one was judged and marked on the weekly record cards secretly reported by the Communist agents. To those who had shown no interest or refused to participate in these competitions, the heavy and dirty work was assigned.

To accept an "old" title, such as Old Teacher, Old Bishop, was feudalistic; to smoke a better brand of cigarette was capitalistic; to stick to one's opinion, whether right or wrong, was imperialistic. Any sign of reluctance and recalcitrance in regard to manual labor and any "smack" of individualism, feudalism, and imperialism, either in words or actions, were noted and reported. Soon the prisoners were classified and transferred to other camps or prisons where military control and forced labor were more rigidly enforced. This was the so-called Reform by Labor.

If a prisoner was fortunate enough to make the qualified grade in the first four stages of indoctrination, namely,

(1) his diligence and progress in studying the Communist-required books, judged by his written answers to the questions prepared on each book by the Self-Salvation Association;

(2) in writing the three required papers, Tzi Pei Hsu (autobiography), Tan Pei Hsu (confessions), and Tzi Sheng Hsu (self-examination);

(3) in passing successfully through the "group criticism" meeting, and

(4) in participating in various kinds of manual labor,

he would be given the fifth and last step of indoctrination, the so-called "Progress Checking Meeting" of the study group, in which the prisoner brought forth was scrutinized again by all the participants. More confessions were expected from him. Comments and criticisms for or against him would be given by the participants, always with concrete illustrations from his daily life—no nonsense or vague assumption would be tolerated in such meetings. Those criticisms of progress were made within the four phases of Communist indoctrination, namely,

(1) his progress made in Communist studies;

(2) his progress made in Communist thinking;

(3) his progress made in Communist living, and

(4) his progress made in being frank and honest in his confessions.

Each phase was divided into three categories: positive progress, some progress, and no progress. These four phases were taken up by the group leader, one by one. After the presentation of criticisms and comments by the participants, with concrete examples, a vote was taken as to which category the prisoner being checked should belong.

One after another, each member of the study group was taken up and checked. Advice was given to the retarded and warnings to the recalcitrant. During the last few weeks, such progress-checking meetings were held and reports of each prisoner were sent in by both the group chairman and the secretary. In the meantime, the secret Communist agents also sent in their weekly reports of each prisoner assigned to them. By comparing the group progress-checking report with that of the secret agents on the same person, the fate of that prisoner would be determined.

Under such group pressure, even the most dumb and stubborn fellow would have learned within a few weeks how to talk "progressively," but to talk "progressively" was not enough. We had to watch every word we spoke and every action, and even our pleasure or displeasure at what we saw or heard. For instance, I had always been called "Bishop" by friends and fellow workers. During the time of confinement most of my fellow prisoners called me "Old Bishop" so as to pay me some respect and to show their appreciation of my medical and other services. (When the adjective *old* is attached to your title, you are respected as a person of wisdom and rich experience.) One of the Communist agents whom I had treated and nursed during his illness, wanting to show his gratitude to me, pulled me aside one night and whispered to me, "You are all being indoctrinated. The sooner you get your past thoughts purged, the earlier you will be released. Sometimes a Communist agent likes to trap you by calling you Bishop or Old Bishop. Bishop is a feudalistic title; Old Bishop is doubly feudalistic. When a person calls you 'Old Bishop,' either don't accept the title or ask him to change it to 'Mr.' or 'Comrade,' or show no sign of pleasure in response, otherwise he may mark down that your thoughts are still feudalistic."

God in His love treated this man and cured him. To God's love, even the stony-hearted Communist agent responded beautifully. "For Thy sake we are killed all the day long; we are accounted as sheep for the slaughter. Nay, in all these things we are more than conquerors through Him that loved us."

16

Communism and Children

ONE EVENING a rich fellow-prisoner named Hsiao, a native of Hupeh, somehow got into the prison, through one of the assistant wardens, two bottles of Mao Tai wine, the famous liquor produced in Kweichow, as well as a pound of roasted peanuts. Our friend, the Communist agent, was invited to the "forgetting sorrow" party. He accepted gladly and acted like a starving tiger looking at his prey. In the midst of eating and drinking freely, someone in our cell became curious and asked openly, "Have you—anyone—heard about the Little Devils' Corps? Is it true that the Communists organize the Little Devils' Corps for the purpose of getting information about their families and neighbors?"

Silence reigned for a few seconds as no one in the cell (of course, Mr. Cheng was an exception) knew much definitely about the Little Devils' Corps, only lots of rumors and hearsay. However, our friend, Mr. Cheng, in a half-drunken mood, was anxious to show off his knowledge and authority, and began voluntarily to lecture to us on the subject. "I know about it," he said. "Let me tell you something. We Communists believe children are sheets of white paper on which we may write what we like, or they are the clay of which we may make any kind of vessel we wish. They can be made the ideal party members, with very strong party-nature [loyalty to Party] by means of indoctrination. That is why children are precious to us and our organization [Party] pays so much attention to the work of the Children's Corps and the Youth Corps. Little Devils' Corps is the name given to the Children's Corps by our enemies. No matter where we go, we are anxious to penetrate into the educational institutions, such as kindergartens, primary schools, middle schools, colleges and, if possible, take them over, particularly the orphanages where the children are without parents and 'burdens' and have had a hard life.

"The members of the Children's Corps are candidates for membership in the Youth Corps, while those of the Youth Corps are candidates for membership in the Party.

"When we take over a kindergarten or an orphanage, we don't change its purpose or its curriculum. The purpose remains fourfold—of developing the child morally, intellectually, physically, and socially. We still have singing, dancing, handwork, stories, and character-learning in the curriculum. They are adapted to the needs of children, but the main difference is in the method of training and the materials used for teaching. Frankly speaking, the object of Communist education is political indoctrination. Children are taught to make a clear-cut demarkation between the people whom they should love and the enemies, such as imperialists, capitalists, feudalists, whom they should hate. Teachers should constantly tell stories of how the imperialists invaded China and imposed on us the unequal treaties; how the capitalists took the wealth away and made China weak; how the foreign sisters and nuns, under the camouflage of religion and charity, killed many orphans and took out their eyeballs and shipped them back to their own countries in order to make telescopes and microscopes, or to be planted to the eyes of their blind; how the Nationalist Party became the enemy of the people; how gloriously the Communist Party has liberated the oppressed and the poor, and how children should grow up and fight against those enemies in order to have abundance and peace. The Communist songs are written to the same effect and taught to the children, and only Communist-approved pictures are allowed to be hung up in the classrooms and orphanage homes.

"The most important work in both kindergartens and orphanages, as in any other organization or institution, is to teach the children how to hold self-criticism and criticism-of-other-children meetings once a week—the proper program is the weekly life-discussion meeting. Every child is told in turn to criticize himself and then others. Often you hear a child getting up and saying, 'When the teacher was facing the blackboard, I hit TiTi once—this is dishonesty,' or another child would confess by saying, 'Yesterday when my little sister fell down in the snow, I walked ahead without paying any attention to her—this is no spirit of mutual assistance.'

"Any child who is able to criticize another child for being disorderly, disobedient, lazy, and so on, is usually rewarded with a red flower or one or two biscuits extra. The children are encouraged to tell the meeting what they have seen, heard, and eaten at home or elsewhere, under the motto: 'Tell all that you know; tell it in the fullest detail.' The teacher, if not a Party member but at least a member of the Youth Corps, usually takes down the report and passes it on to the Party.

"If the child reports that he has eaten such and such nice dishes at home, the Party will put heavier taxes on the parents or ask the parents to buy more bonds or make bigger contributions to help the war front. If he reports that such and such persons, known or unknown, have come to see his parents several times, or that they have listened to the Voice of America or the BBC, they are arrested immediately for questioning and investigation. Any information which leads to the arrest of any person, sometimes likely the child's own parents, is rewarded in various ways. This function of the children at the weekly life-discussion meetings is very valuable to the Party. It is because of this function that the public has the impression that the Little Devils' Corps is organized for the purpose of spying on their parents at home and on others outside."

17

Class Classification—Our Headache

WHEN WE OPENED *The New Democracy*, by Mao Tze-tung, the bible of the Chinese Communist Party, the first lesson we learned was about the human class-nature. In society there are classes and each class has its definite characteristics.

In the old society, five classes of people were traditionally recognized: (1) scholars, (2) farmers, (3) laborers, (4) merchants, and (5) soldiers. Their degree of importance was according to this sequence, as two Chinese popular sayings have well expressed it: (1) "All vocations are mean but the scholar is the highest," and (2) "Nails are not made of good iron nor soldiers of good children." These sayings also indicate the deep-rooted conception of classes in society in the minds of the Chinese race.

Both the Nationalist and the Communist parties, in order to promote patriotism, have reversed the order and changed the latter saying into "Good nails are made of good iron, so good soldiers of good children." As this harmonizes with the current of patriotism, it seems very natural for the masses of people to be taken in by the Communists' propagandizing about classes and human class-nature. Furthermore, in Chinese history, the people have always enjoyed the freedom of changing their vocations from one to another at will. Many great emperors and renowned statesmen in the past came from humble origins, being children of poor farmers or small merchants. Lately, in the last war, many professors changed from teaching to business. There has never been a caste system in China as there is in India. Consequently, when we read about classes and class-nature, no one had any feeling of doubt, disagreement, or opposition.

No sooner had recognition of classes in society been made than we automatically came to the problem of how to divide the people into classes and sub-classes, to distinguish one class from another. To learn about the classes and their characteristics in general was easy, but to understand how the Communists had arrived at the sub-classes of various major classes was a headache, the simple fact being that they gave nowhere any definite standard by which we might subdivide the classes and differentiate between one class and another.

Before the Communists were in power, they had spoken very vaguely in their propaganda of only two main classes in society; those in power were the oppressors, and those not, the oppressed. The oppressors were the exploiters while the oppressed were the exploited. With the banner of emancipation for the oppressed hoisted high in the air, they easily obtained lots of sympathy from the masses, but after they came into power, they were obliged to divide and sub-divide the oppressed.

As indicated on the Chinese Communist flag, the four small yellow stars following the lead of a big yellow star (the Communist Party) really represented the four major classes of the oppressed people, as follows: (1) the laborers, and particularly the producing laborer who owns no property whatsoever, are the real and only proletarians, full of the spirit of revolution; (2) the farmers, who had been shedding blood in the revolution, could be counted only half proletarian because, usually, they were conservative; (3) the bourgeois, including scholars and students, who, for their tendency to make compromises easily, must be regenerated (brain washed), and (4) the national capitalists, who had given lots of material assistance to the revolution, were to be utilized as much and as long as possible but in the long run would be liquidated. In other places, they defined "State" as consisting of only three classes of people, *i.e.*, the policeman, the soldier, and the prisoner.

I remember one day when we were required to write our confessions (Tan Pei Hsu), in which there was an item, "Classify the class of your parents, including their vocations and thoughts of the class to which they had belonged," we all had a hard time writing down the class or classes of our parents. Most of our parents were farmers. According to the Communist books, farmers are divided into four major classes: (1) the landlord, (2) the rich farmer, (3) the well-to-do farmer, and (4) the poor farmer. Each major class is again subdivided into three more minor classes, so that there are at least twelve classes of farmers.

At first we all knew that the landlord was the sinner. Pretty soon we learned that the rich farmer belonged to the condemned. Only the poor farmer was the chosen one, while the well-to-do farmer was declared to be "separated" and "protected." None of our fellow-prisoners had parents who belonged to the poor farmer class. By poor farmer, the Communist meant farmers hired day by day, or annually, or tenant farmers without any property of their own. As a result, to classify our parents as well-to-do farmers was the only way out, but to which sub-class of the well-to-do farmers?

By what standard? Amount of land? How many *mow*? What are the regulations?

All these questions came up in our minds. For days we talked about it and discussed it in our study groups. Arriving at no definite answer, some of us dug into the Communist books on land reform but still failed to find anything definite except one statement which says, "If the income of the family members from labor is more than 75% of the total annual income, that family may be classified as the well-to-do farmer." Even that statement was vague! What do they mean by labor—only manual labor? Who decides the total income? The actual income? The estimated income?

During our discussions, one prisoner, who had escaped from the Northeast but had been caught in the Southwest, related the story of an old Confucian scholar who, by means of tutoring, had accumulated about 20 *mow* of land (six *mow* is equal to one acre), a part of which he let out to a tenant farmer while he continued teaching and tilled the rest of his land himself as a hobby. During the Land Reform in the Northeast, he was considered neither a rich farmer nor a well-to-do farmer but a landlord, and was tried, condemned, and liquidated.

Another prisoner followed with a story about a girl he knew, a native of North China, who was very poor but progressive. Soon she was admitted as a candidate for Party membership because she was a proletarian. However, it was later found out by the Party that her family had one water-buffalo for rent and her status was immediately changed from the proletarian to the rural capitalist. Because of the change in class, her candidacy was automatically dropped.

In the midst of this confusion, we therefore appealed to the prison authorities to get a Communist theorist to come in and clear up these points so that we might be able to complete our confessions. They graciously consented. One week later, a man a little more than thirty years of age, in Lenin uniform, did come and lectured to us about forty minutes, then he happily and proudly gave us permission to ask him any questions on class-classification. He acted as though he knew everything and spoke as if with great authority, but in his lecture it was the same old story that we had read in the Communist books. What he gave us was just a few principles, lots of class names, and the nature of each major class. As to the core of the problem—the standard by which we could classify classes and sub-classes—he did not touch this at all. However, during the question period we had lots of fun and gained some insight to the problem.

One well-educated instructor, intending to put our theorist on the horns of a dilemma, started the ball rolling by saying very politely, "Our new government is founded, as we have learned, under the leadership of the laborer and on the foundation of the united forces of the laborer and the farmer. The Communist Party is the party of the laborer. How can you classify Chairman Mao as a laborer? We all know he was an assistant librarian at Peking University and is a great scholar!"

After thinking for a moment, our theorist answered, "We don't mind a person's background, whether capitalist or bourgeois. As soon as he joins the laborer's party, he rebels against his former class and becomes a laborer."

Our instructor immediately fired another question, "Does this mean that all the laborers are not necessarily revolutionaries, but that all the revolutionaries are laborers?" The theorist turned his face aside and kept quiet, not saying a word.

Another prisoner, having waited impatiently, finally grasped the chance and asked him, "Comrade, please give us the standard by which we may distinguish one class from another."

Looking at him a little angrily, our theorist became very authoritative and said, "You have no right to call me 'Comrade' because you are a prisoner and reactionary. Not until you have been regenerated can you do so. When you ask for a definite standard you are only proving that you are not regenerated. My advice to you is to study more and digest more Marxism, Leninism, and the New Democracy. In due time you will grasp the principles; those principles will enable you to think and act properly, and to distinguish classes. We have only principles, not laws or regulations. They are things of the reactionary—such and such section, such and such number. So many regulations! So many bondages! The people are just slaves yet the reactionary talks so much about freedom—this freedom, that freedom! In fact, there is no freedom at all!"

The Communist hates the idea of freedom. It is his duty to prove that the so-called freedoms in the democratic countries are just illusions. Everything is conditioned by its cause or causes or environment. Our theorist, seeming to have scored a great victory in speaking against freedom, was quickly changed. He was happy again. Words and sentences flowed from his mouth rapidly without any thinking. In conclusion, he said emphatically again and again, "When you are regenerated, you will know the principles; those principles will give you real freedom."

It is true, in a way, that we have too many laws and regulations in the democratic countries and it is so in the Church with our canon law. Time and again some people break the law, sometimes without knowing it. Nevertheless, we still have freedom within the law and regulations.

Now it was all clear except one main question: what is the real intention or purpose of the Communists in keeping principles and abolishing laws and regulations? It was only after much careful study, keen observation, and listening to the Communist lecturers on various occasions making such statements as, "The strategy of war is the same but value lies in how to use the strategy appropriately in accordance with time, space, and situation," that our headache was finally relieved. The purpose of giving the people only principles, not regulations, is to keep all freedom in the hands of the Party in the interpretation of those principles. When in need of popular support, their interpretations were liberal, while for liquidation they were rigid.

Thus, in such a way, the class of any person is determined. It has to be studied and decided by the class-classification committee and approved by the authority of the local party organization. With only principles, the determination is largely dependent on the impression of the committeemen at the meeting, the will of its chairman, and the mood of the local party leader. Principles may be twisted but not detailed regulations! Truly, "Children of men are wiser than the children of God!"

18

My Trial at Night:
How Truth Nearly Killed, but Saved Me

DURING THE TIME of indoctrination we learned that one of the Communist doctrines which attracted as well as fooled so many readers was "Service to the people is the object of life." This almost exactly parallels the Christian teaching, "To minister, not to be ministered unto."

In the old Chinese culture, the *family* was the unit of duty, faithfulness, and obligations. Its scope might be enlarged but in practice it was always limited to those whom they knew and with whom they had some close relationship, as brethren by blood covenant, members of the same clan or of the same secret society, graduates from the same school or college, natives of the same province or district. It never took in every person in the whole nation, in spite of the Confucian popular teaching, "Within the four seas are brothers."

Having no concept of the nation in comparison with the family and friendship was undoubtedly one of the causes of China's weakness, the weakness of putting too much emphasis on the family, relatives, and friends instead of on the nation, yet her weakness was also her strength. During the Second World War, we were driven step by step from the coasts by the Japanese. Thousands and thousands migrated to the mountainous provinces in West or Southwest China.

Many of them, though poor, walked day after day, week
after week, even month after month, with their belongings
on their backs, to the hinterland of freedom. This was not
because the civil or the Christian organizations were able to
cope with the situation by means of providing sufficient
relief, or because the government had set up assisting stations,
but merely because of faith in a relative, a friend, a school-
mate, a blood-covenant brother, or even an acquaintance who
had established himself there. It was the duty of these people
to help, and they did help. I haven't heard of any refugee
ever being turned down by his friends or relatives no matter
how poor or how large the family he had with him. This
weakness of China was one of the elements that enabled her
to stand against the onslaught of the Japanese during those
eight long years of war. Thus, to a great extent, her weakness
was her strength.

The Communists at one time captured practically all the
idealistic youth, the educated, and the native Christians,
capitalizing on this weakness and providing the "cure" of
service to the people in order to make China strong as a
nation in the way the Nationalist party had done before,
only to a greater degree. I doubled my efforts in helping our
fellow-prisoners, first by explaining and solving the problems
raised from the course of the Communist indoctrination. In
addition to answering questions in connection with New
Democracy and Communism, I sometimes had to take a
whole book and analyze it; at other times I even had to write
book-reports and confession papers for the illiterate. "To
minister, not to be ministered unto!" "Service to the people
is the object of life!" A great motto! A great slogan! Magical
and mysterious! Often I had to work sixteen hours a day in
order to fulfill my service to others!

Imprisoned with us was a Mr. Chen, young and active. He was a clerk and a leader of the younger workers at the Kunming Mill. Under his leadership various athletic teams were organized. Besides, he liked to hunt and for the purpose of hunting—and self-protection—he owned a shotgun and a small pistol. During the last two years before the Communists took over the mainland, he was approached several times by the underground workers. As an athlete, busy with his teams, young and gay, he had no taste for politics and showed no interest in the Communist requests. For his refusal to co-operate, he was thrown into prison immediately after the underground had taken over the city on a charge, as he later learned indirectly, that he had a pistol. Confidentially, he told me that he still had a chance if he would surrender his pistol and make progress in his studies. Therefore, he had to make the grade and straighten out his thinking so as to redeem himself.

One afternoon he came to me with a philosophical problem, asking, "Old Bishop, will you please explain Truth to us? I have read and read, thought and thought, but to save my life, I cannot make out anything from the definition. What do they mean by 'relative'? What do they mean by 'the people,' which occurs so many times in the definition— as 'approved by the people, experimented among the people, and for the benefit of the people'? It is really beyond me. Please help!"

Putting other questions aside for the time being, I started to explain to the group of five what the Communists mean by "truth." Their definition of truth in one place states: "Truth is that which is approved by the people, experimented among the people, and always for the benefit of the people." In another place, as Mr. Chen said, it is "relative." Going directly to the subject and trying to simplify the mystical phrase, "the people," I said, "It may be paraphrased like this: truth is some thing, such as a project, a plan, or an undertaking, which must first of all be approved by the Communist Party. This is the first 'people.' By 'experimented among the people' means that that project or undertaking should be tested out among the masses of people. This is the second 'people.' Again, 'for the benefit of the people' means no more and no less than for the benefit of the Party-State, the sole purpose of all actions. This is the third 'people.' "

With a sigh of relief, Mr. Chen declared, "You have made it so plain and so clear to us. Thank you, Old Bishop! In another book it says, 'Truth is relative.' What do they mean by 'relative'?"

"Here," I said, "is a great point of difference! According to various systems of philosophy and schools of theology, truth is rather universally defined as *absolute*. Truth is consistency. Everything changes, but not truth. To the Communists it is entirely different; truth changes as does everything else. This is what they mean by 'relative.' They claim that only the Communists know and have truth. Truth varies according to space and time. For instance, the Communists speak against colonialism in the colonies and against semi-colonialism in China, but never in Russia. This is change according to space. Before they took over the government, they had constantly protested against arrests and advocated personal freedom, but now here *we* are. This is relative according to time."

It was wonderful to see the change in their facial expressions and their appreciation of my plain, direct explanation! When we broke up, I jokingly made one comment: "We really can improve the definition by saying that truth is not only relative, but also absolute. It is a pity that they have not done so themselves!"

Unexpectedly, one of the five at the meeting reported the whole explanation of truth to the Communist authorities. They did not like my explanation at all as being "too bare to the bone," and particularly my last casual comment. Later I heard that two meetings of the Communist theorists were called to discuss what to do with me. Someone in the meeting even declared, "Who is he, anyway? A prisoner and a reactionary! Tries to improve our definition! There must be some ulterior motive. We must liquidate him or at least humiliate him and bring him into line!"

Consequently, they decided that some sort of people's trial would be held soon at night in the room just outside the guard room, but still within the prison walls, and that I should be brought before them, tried, humiliated, and brought into line, otherwise I would be transferred to another prison. They also decided not to call all the prisoners to the people's trial because I was then pretty popular among them, due to my various untiring services to them.

In the meantime, while they were planning for the People's Court, my Communist friend, imprisoned with us as one of their agents, quietly and confidentially warned me of the coming trial, advising me, "Bow down before the people and confess your faults. This is the best and only way out; otherwise, I cannot think of the consequences!"

One night at 11 o'clock, when every cell was locked up and every prisoner in the dream-land, our cell was opened and in came two guards calling my name. Somehow, that night, I could not sleep at all, but just changed from one side to another even though totally exhausted. I got up immediately, pretending to be very calm although, in fact, my heart was jumping up and down inside. Usually trials held at night were considered serious. Fortunately, the guards were quite friendly and did not rush or blindfold me.

Soon I was walking between them, followed by my Communist friend, shivering in the bitter cold winter night. The moon was shining over us in the small courtyard but its brightness seemed to have lost the romantic touch. Looking at her, I bowed my head and in my mind I prayed, "O God, Thou are almighty and merciful! Give me wisdom to say what I ought to say, strength to stand what I have to go through, and faith in Thee, so I may not bring shame to Thy name. Amen." After that prayer, I felt much better, greatly strengthened, and peaceful in mind. Both my fatigue and exhaustion were gone!

On entering the court room I saw three men sitting behind a long table, as judges, in one end of the room, and a dozen or so others as the audience or the people. The light was dim and I could not distinguish them. They were all in the Lenin uniform. The air was heavy and the situation tense. Here, I heard one voice calling out, "Here comes the reactionary, a spy. Get rid of him tonight!" There, I heard, "Under the cloak of religion he worked for the imperialist—a running dog of American imperialism. Kill him!"

A few minutes later the judge in the center called the trial to order and the judge on the left started the trial by asking my name, age, birthplace and all my past history. Once in a while I was interrupted and asked to give more information or to repeat some points, or to tell them this connection and that connection—with American friends and organizations, with some leaders of the Nationalist government, with the Rotary Club, with other denominations, etc. It took me at least an hour and a half to complete my story.

Before I concluded, the judge on the right commanded, in a harsh voice, "Now answer my questions. Did you explain truth to them the other day? Are you sure that your explanation is correct?"

"Yes," I answered, "I did, because I was requested to by Mr. Chen. I did it because I was trying to fulfill the purpose of life—service to the people. Of my explanation, right or wrong, I am not sure. I shall be grateful if I may have a chance of being corrected. I explained it plainly and directly, just as a teacher to a student, in order to make them understand. Furthermore, I was following one of your principles, 'We should always tell what we know and tell it thoroughly.' If I made mistakes in helping them or committed faults in explaining to them, I shall be glad to know and make corrections. My understanding may be wrong but I cannot help it, as we are all our own teachers. We have petitioned again and again for some authority to come in and teach us. Only once was our plea granted."

In the midst of the audience came a sharp voice, "Clever argument! Bow down before the people! Confess! Repent! Otherwise . . ."

Following my defense, the judge in the center began to question me in a soft, friendly voice, "Well, we see now why you so explained, but you must admit you did it a little too much. Possibly you might have given them a wrong impression. We forgive you this time on this point but be careful. Say, how do you explain that you can improve our definition by saying that truth should be not only relative but also absolute?"

I answered, "Judge, I don't think I committed any fault by saying it is relative." "Oh no," he answered, and immediately asked, "How about absolute?" "Yes sir," I answered,

"I did say we can improve the definition by adding the word 'absolute.'"

Without waiting for the completion of my answer, a voice from the crowd shouted out, "Absolute! Who taught you that? Damn you!"

"Please let me finish," I pleaded, "condemn me, if you wish, *after* I have completed my explanation. By absolute I mean that no matter under what condition, at any time and in any place, the unchangeable purpose of any action is for the benefit of the Party-State. This is consistent—only by this we may have a strong and prosperous China."

By the patriotic element in my answer (that was what was being promoted far and wide throughout the nation), our half-baked judges and theorists seemed more or less moved, and all became quiet except for one or two among the people still calling out, "A reactionary, a spy, a running dog of American imperialism! Damn him! Kill him!" The head judge in the center got up and, after scratching his head a little with his right hand, said, "Let us conclude our trial tonight; it is late. We may call another trial later."

Then I was again escorted back to my cell by the same guards. Lying on the floor, with my head against the wall, I was still unable to sleep but my heart and mind were full of gratitude and praises to the Lord, for He had given me wisdom and courage to say what I ought to say. While I was offering my prayer of thanksgiving, Mr. Hsiao, lying on my right, whispered in my ear, "I am glad to see you back safe and sound. We have worried about you and were afraid that we would not see you again!"

For the next two days I was sleepless and speechless, constantly thinking and preparing myself for what I would say to defend myself in the next trial, as declared. Time and again I was anxious, worried and nervous; also, time and again I was comforted by recalling the words of our Lord to His Apostles, "But when they deliver you, be not anxious how or what ye shall speak, for it shall be given you in that hour what ye shall speak. For it is not ye that speak, but the Spirit of your Father that speaketh in you."

Two days later, to my great surprise, a message was given to me through a Communist agent, saying, "There will be no more trials; you are asked to give lectures on New Democracy and Communism to a group of prisoners. Be careful of what you say beyond the books!"

Thus, my humiliation, disappearance, or death was turned, by the grace of God, into my opportunity. After that, I was considered by my fellow prisoners one of the authorities on New Democracy and Communism!

19

Subtlety of Communism

ONE EVENING at 5:30, after we had had our second meal of the usual fermented rice and "glass soup," an assistant warden came and told Mr. Ho Tao-Chih, one of our eleven cellmates, that he had been given permission to go to the gatekeeper's room to see two Liberation Army comrades waiting there for him. Ho was surprised and could not figure out who they were, yet felt very happy and much honored by a visit from the two comrades. He hurriedly cleaned himself up a little and hopped and jumped through the crowd. A few minutes later he was face to face with his lost younger brother and an unknown companion. After a forty-minute visit he returned happy and bewildered; happy because his younger brother lost for two years was now found, but bewildered because his brother presented a difficult problem to him.

Seeking my advice, he confidentially told me the story of his lost brother and the problem, saying: "Our native district is on the border of Indo-China where my father, in addition to what he had inherited, bought hundreds of *mow* of rice fields and farming land. As he had done so much for the people, at his death we served hundreds of tables of funeral dinners for forty-nine days so the people of the district could pay their last tribute and respects to him (Note: forty-nine days means 7 days times 7, the round number and the period of the Buddhist funeral service by which the soul of the dead may be redeemed from condemnation and go to Nirvana—the Buddhist paradise.)

"After his death I did not want to be shut up in our home district and have been all over the country trying to help build a prosperous and strong China. I married outside and our younger sister later joined us while our widowed mother, so anxious to keep the family properties, remained at home and took care of her beloved son, my younger brother, Li-Chih. But in the midst of the last war, my home town became one of the strongholds of the Communists under Chuang Tien. He was trained both in Yenan and Moscow. The Communist force was getting bigger and stronger but, fortunately, they did not molest the people. On the other hand, there were also lots of bandits who regularly plundered the peace-loving farmers and merchants.

"One night two years ago a score of bandits, with their faces painted and heads covered, ransacked our old home for five hours. They carried various kinds of weapons—swords, spears, rifles, and pistols. My mother was locked up in the attic and was so frightened that she fainted and lay on the floor for hours. My brother, though young and strong, was only fifteen and could not fight against them. Soon he was tied down to a *hwa ken* [*Hwa ken* is a means of transportation used in the mountainous provinces. It is made of two bamboo poles about ten feet long, between which a piece of flat bamboo is used as a footstool and a few more as a seat, and the back is tied with ropes, making a sort of primitive chair for the passenger to sit in. It is carried by two or four coolies.] After having ransacked the place, twelve of them carried away the booty, including all our family valuables, clothes, and bedding, and two carried away my younger brother, tied on the *hwa ken*. The remaining six served as guards, marching off in the dark and quiet of the night. Up hill and down dale, step by step, they slowly covered twenty-five *li* [a li is equal to ⅓ of a mile] in four hours. Just before dawn they came to a crossroad, intending to have a short rest.

"Suddenly there was a thunderous noise all around and a particularly sharp voice yelled, 'Who are you? Stop; otherwise we'll shoot.' Outnumbered three to one and with machine guns pointing at them, they quickly and quietly surrendered themselves, with their booty and Li-Chih, without firing a shot. This fulfilled one of the Chinese sayings: 'Thieves have met robbers,' with the bandits meeting the Communist soldiers. After daybreak, the bandits were disarmed and disbanded, while the Communists took the booty and the prey (Li-Chih) to their headquarters on top of the Southern Mountain. Thus Li-Chih was saved by the Communists, but lost, because after that we did not hear from or about him.

"Li-Chih felt deeply grateful and, to express his gratitude for being saved from the hands of the bandits, he voluntarily begged to stay and work with the Communists for the cause of the revolution. During the past two years my brother risked his life many times in order to fulfill the missions ordered by his superiors. As a result of his loyalty and daringness he has been granted full membership in the Communist Youth Corps. During the next six months, if he can earn three more big merits or nine small merits, he will be qualified for full membership in the Communist Party. He is certainly an ardent believer in New Democracy and Communism and always carries a mimeographed copy of the book, *The Directory of Thought,* in his hand. Look at it; I have borrowed it from him. He has been studying New Democracy and Communism very diligently and finding every opportunity to gain the necessary merits.

"He has talked with my wife and sister four times and urged my younger sister to become a candidate for the Communist Youth Corps but they insisted that my approval must be secured first. That is why I was so honored by the visit from him and his companion. I was not able to refuse the request in front of his unknown comrade but told him that I would give careful and serious consideration to the problem. But, in my heart, I have made up my mind to send my younger sister to your Hueitien Hospital [the Episcopal Church hospital in Kunming] to be a trained nurse. What do you think? What would you advise me to say and to do when he comes back again in a few days to get my final answer?"

Immediately my mind became alert and critical, and I asked him a number of questions: "Are you sure that your younger brother was saved by the Communists or *fooled* by them? Did the bandits and the Communist soldiers—one group to loot and kill and the other to save and protect—belong to two different camps or just to one camp, like two different actors in the same play?" About his sister, I commented, "Whether or not you will give your approval, to them it means nothing. The head of the family is out-dated and has no more authority over its members. The best you can do, in order to comfort your own conscience a little, is to insist, even without effect, that she be given a chance of nurse-training so that she may serve the cause better and more tangibly. Your brother needs this merit to qualify him for Party membership and the Party needs young girls, as we have all heard and learned. Your approval or disapproval will make no difference."

With these questions and comments of mine, Mr. Ho Tao-Chih thought and pondered over the problem. The more he thought, the more suspicious and confused he became. He worried, he could not sleep, and he was a little afraid to see his brother again, yet he was anxious to find out the truth, whether he had been saved or fooled by the subtlety of the Communists.

Four days later his younger brother came to our cell, this time all by himself, with the definite purpose of getting approval from his elder brother since the sister had promised to join as soon as approval was secured. He was then introduced to all of us in the cell but he did not pay much attention to us—the reactionary prisoners. Starting off the conversation with his brother, he said, "I am sure your answer is YES. It cannot be NO. Yes is good for us all—good for you, to prove that you have progressed in your thinking and that progress will redeem you; good for me, in that I shall be credited with a big merit; good for sister, as by this first step she may build a great future for herself."

In spite of his dialectical thinking and speaking, the elder brother quietly and slowly said, "Before I give you my approval, I would like to find out one fact—the truth about yourself. You told me the story the other day in great detail about how our home was looted by the bandits and how you were saved by the Communist soldiers. It was wonderful—but have you ever during the last two years met anyone or heard anything that would be a clue for you to suspect that the bandits were, after all, also Communists?"

Hearing this question of doubt, Li-Chih became a little peeved and, in a rather angry tone, scolded his elder brother, saying, "What is the matter with you, elder brother? You are still holding on to your bourgeois thinking and suspecting of others. I was *saved* by them, otherwise I would have been killed by those bandits. The Communists were my saviors; I have been feeling ever so grateful to them."

After having said these brave words and other expressions of loyalty and gratitude, he was hit, and hit hard, by something in his mind. Suddenly he paused and dropped his head. He did not say a word and did not care to listen to his brother, but sat thinking for the next few minutes. At last, in a very small voice and with his mouth close to his brother's right ear, he whispered, "Perhaps you are right, because under the dim light of our oil lamp I could recognize a little the person who tied me to the *hwa ken* and, furthermore, I could recognize his voice very well. He seemed to be the leader, directing and giving out orders to the others. After I had had three months' indoctrination at a separate camp, I was called in and put to work in my boss's office. Again and again the assistant boss appeared in our office. I have never suspected him but always thought I had met him somewhere before. His face and voice are so familiar to me!"

20

A Student for the Ministry Who Worked for the Devil

BEFORE I took over the episcopate in 1946, I had been informed about the only theological student of the missionary district of Yunkwei (which was changed into a diocese in the fall of 1947), who was taking a combination Arts and Theology course at Hua Chung University, located then in Hsichow, Western Yunnan. His name was Ho Ping Chung, a native of Kunming and a nice young fellow, whom I had met once in the summer of 1945 when he was helping with our church work at Wen Lin T'ang (a student chapel with various activities for the students of the Associated University of Southwest, University of Yunnan, Kunming Teachers' College and Wuhua College). After World War II, he followed Hua Chung University to Wuchang, Hupeh, and remained there as a student, while I came over to the States and was consecrated bishop for that district on August 14, 1946 in Santa Barbara, California. After that, I was constantly in touch with him and encouraged him with great hopes for the work of God in Southwest China and his responsibilities as a native of Kunming. Occasionally I sent him bundles of religious books that I had collected from friends in the States. Because of the strong provincialism in Southwest China, particularly in Yunnan Province, I stressed a great deal the importance of the natives joining the ministry. I did all I could for him, with a fond hope that some day the whole Church of God in the Southwest would be headed by natives.

In the spring of 1949 he returned to Kunming. One day he suddenly appeared at our house to see me. I was anxious to know his story and asked him, "Have you finished Hua Chung University and gotten your bachelor degree?"

"No," he replied, "I have not finished but have only half a year more, though. That is just what I came over to explain to you today. When I was at Hua Chung in the beginning of my senior year, Dr. Wei studied my records and sent them over to the Ministry of Education. Correspondence was carried on then for a term between our college and the Ministry of Education, and the final conclusion was that I had to take more courses—at least one extra term—in order to get my bachelor degree. By the end of the first term of my senior year, I was very much discouraged and, at the same time, I heard that my mother was very sick. In a few hours I decided to go back home, and started my journey. It happened so suddenly that I did not have time to let you know beforehand. I am sorry. Now my mother is much better and I intend to find a teaching job for a year or so, until the present war conditions get settled down."

I was very sorry when I heard his story, together with his excuses, and still hoped that he would go back to Hua Chung University as soon as possible, so I said, "I am sorry that you have come back without finishing your college education, but I am glad your mother is better. You know very well that our diocese is autonomous and independent. To the glory of God, we aim at making the Church here not merely of the Chinese but also of the natives born in Yunnan. In you I have great hopes and as soon as you can take over, we shall

go back home in Central China. Since you have given your
life to the work of God and have had so much trouble with
the Ministry of Education, let us forget your B.A. degree for
the time being, but let us do our best to get your B.D. degree
from Hua Chung. I'll write a nice letter to President Wei,
asking him to make arrangements for you to finish your
theological training. Now get ready and go back as soon as
possible. I'll provide the necessary funds, including enough
for your travel to Wuchang."

At the end of my talk, he showed deep gratitude but
argued that conditions were very much upset along the
Yangtze River, as the Communists were preparing to cross
the river and ready to take Shanghai, and that it would be
better for him to wait for a term or two. However, I still
urged him to get back in time for the opening of the second
semester but his argument was stronger in the face of reality
so we decided to leave the matter for the time being, and at
last he departed happily.

After that, I became busier and busier every day, traveling
through the diocese, planning and organizing self-supporting
projects and emergency measures, holding sectional emergency meetings and the Diocesan Emergency Synod, as the
Communists were sweeping over, first, East China, then
Central China and South China, and approaching Southwest
China. In the meantime, Mr. Ho was supposed to be teaching
at a suburban middle school but he never came to see me, and
no chance brought us together for a year or so.

On the afternoon of March 12, my name was called again.
This time I was told that the judge was a Mr. Tao. Led by a
guard, I entered the small room opposite the prison inner
door. Judge Tao got up and very politely greeted me. He

closed the door himself and placed a chair for me to sit down on the opposite side of his desk. In response to his utter friendliness and courtesy, I was very polite, too, as when strangers meet for the first time. This was my third afternoon trial, but a trial in name only as it became a congenial conversation, in fact, when he started by asking me, "Do you remember me?"

Trying to recall, I replied very slowly, "Judge Tao, I am afraid . . . I don't, as I am physically so weak and mentally so dumb. I cannot recall at all; however, I do believe we have met before, though."

"I am Ho Ping Chung, your theological student," he said.

"Oh, Mr. Ho! But why are you called Judge Tao?" I inquired, happily surprised.

"Well, you know . . . in the course of a revolution we seldom use our real names," he explained.

"I am certainly glad to meet you here today," I said, "to say nothing of any other relationship between us, at least we are friends. For the sake of our friendship, will you please tell me the charges against me and the names of my accusers? I am very anxious to know."

He replied in a very soft voice, "Old Teacher, don't be bothered about your accusers. Between you and me, I may let you know that there are three charges against you. First, you are an American spy, for three reasons which are all true: because during World War II both you and your wife entertained the American GI's almost every day in Kweiyang [It was true that we entertained them so much in Kweiyang, our GI friends called my wife "Chinese Mom"]; because you are the only Chinese bishop consecrated in America, and because you do have lots of letters from America. If you don't have any special relationship with America, why were you not consecrated in China as other bishops? Secondly, you are also a spy of the Nationalist government. Thirdly, you are charged that with the funds of the Hueitien Hospital you have sent a few Church workers to the States to be trained as spies against the new regime. However, the second and third charges have been thoroughly investigated and now are cleared. You don't have to worry about them. Nevertheless, the first charge, that you are an American spy, is very serious, and those three reasons are true—you know that."

"For heaven's sake," I exclaimed, "I could be shot for any one of the three charges. You know I have never been a spy and don't know a thing about spying. If I were accused of preaching the Gospel, I would readily admit it, but as to being a spy for anybody, I will never confess. I have never had any interest in politics and have warned my fellow workers not to talk about politics. When we were at war against Japan, so were the Communists, and the Americans came over to help us. As patriots, both my wife and I felt, and still feel, that it was our duty to be friendly with and help our helpers. There is no reason for my being an American spy.

"As to my consecration in the States, this was the decision of the House of Bishops in China and not *my* will. I was consecrated in Santa Barbara, California, on August 14, 1946, and am the only Chinese bishop of the Holy Catholic Church in China to have been consecrated in the States. I cannot be blamed for that.

"I *do* have lots of friends in the States with whom I correspond. They are Church people and not one of them is an official in the American government. They have never asked me to do, nor have I done anything in politics. Those 'reasons' are true but that cannot prove that I am an American spy."

"Well," he thought for a while, and argued, "those reasons may not be valid enough to prove that you are an American spy; nevertheless, you cannot prove that you are *not* an American spy."

His statement was true and I could not prove that I was not an American spy because we were all cut off then from connections in the States by the "curtain."

"Although that charge is very serious," he continued, "you don't have to worry too much, because you have a student and friend who can help you—provided that hereafter you are willing to serve the people. You understand what New Democracy means. You can do it, we know. There is no use for us, especially Christians, to talk about life beyond. Be practical; let us talk about and improve this life here. There is no sense to talk about and preach the Kingdom of God on earth. Let us build it—what we mean by the Classless Utopia. Look at the history of the Church. Almost two thousand years have gone. Where is the Kingdom of God on earth? What have the Christians done toward the building of such a kingdom—when you consider their divisions, jealousy of one another, inquisitions, exploitation by means of unequal treaties, sale of opium to China by force of arms, educational institutions which lead the youth of China to false and impractical idealism and spiritualism, racial prejudices, unfair treatment of the native workers, or causing them to be 'yes men' or 'running dogs' of foreign missionaries? Thousands of years have gone by; millions of years will not be enough! There will never be a Kingdom of God on earth.

"On the other hand, look at the history of Communism. After World War I, came the existence of the Socialistic

State of Soviet Russia, and after World War II have come those Socialistic States in East Europe. Again, look at the Communist Party in China. It was first organized in 1921 and now is sweeping over the mainland of China. Pretty soon there will be a Socialistic State in China! Pretty soon again there will be the Classless Utopia, not only in Soviet Russia and China, but also in the world! How glorious is the history of Communism! How glorious your contribution would be! How great you would become, if you decide to serve the people!"

While he was talking and I was listening, I wished I could say to him, as our Lord had said, "Get thee hence, Satan; for it is written: thou shalt worship the Lord thy God, and Him only shalt thou serve." I was amazed at the cleverness of the devil in capturing the minds of the Chinese modern youth, even that of a theological student, by preaching the gospel of *quick* building of the Classless Utopia in contrast to the *patient* building of the Kingdom of God on earth, by utilizing the Chinese pragmatism taught by Confucianism, which limits everything to this life, and by emphasizing the weakness and faults of Western missionaries and Christians to overshadow the good they have done.

At the end of his exposition of the devil's gospel, I said, "As to the question of serving the people, my behavior both before and after my imprisonment has answered that. To be real Christians we ought to, and must, serve the people, as our Lord has taught us, 'not to be ministered unto, but to minister.' But I don't want to be great. Our Lord has said, 'Whosoever will be great among you, let him be your minister; and whosoever will be chief among you, let him be your servant.' I prefer to be a servant and 'small potato.' As the Bishop of Yunkwei, I have been having so many headaches, more than enough headaches! Let me remain a 'small potato' and live a life of peace!"

He seemed a little bit impatient and asked me, "Have you decided to serve the people, Old Teacher? That is the only question I would like to have answered."

"Yes," I answered, "my decision to serve the people is as old as the Communist Party [I decided in 1921 to study for the ministry], otherwise, I would not be working in the Church."

He exclaimed, "I am awfully glad that you have decided to serve the people. I am sure you can do a lot, and you will be great. I'll see that you get released. Just be patient a few days more!"

So we departed, and I went back to my cell. For days, my mind was filled with those magic phrases, "a few days more," "to serve the people," and "to become great!" I wished I could have a Communist dictionary, to find out exactly what they really meant by those phrases. Temptation, doubt, and fear wandered back and forth in my mind; I had no peace! I was perplexed and had no desire to talk, and I would lie in the quiet corner of my cell, thinking, praying, and at last sleeping.

For the first time in many days, we did not have the group discussion or criticism meeting one evening. When the cell was locked and bolted at 8:30 p.m., I was awakened by my cell mates and told to go back to my allotted place between my Christian companion on the right and a Communist agent on the left, whom God had treated and I had nursed. After I had crawled into my bedding roll, my Communist friend showed his concern for me and whispered, "I am glad that you have decided to serve the people. You will be released in a few days, and you will be great—the great leader of the Religious Revolution in Southwest China. Congratulations!"

Very much surprised at the speed with which he got the news, I said, "You fellows always like to be mysterious and use language beyond my understanding. As friend to friend, will you please tell me frankly what you really mean by 'service to the people' and by 'becoming great'? To serve others is one of the objects of my life, otherwise you would not have been nursed nor others helped by me in this jail. Shall I be permitted to do what I can for others, or do I have to do certain things, whether I can do them or not? It seems strange that I, who have been doing so much for others, should be asked to make such a decision. Another thing which has bothered me is that I shall become great, and the great leader of the Religious Revolution. I don't want to be great! To be what I am has caused me enough headaches. I don't want more—to kill myself. In religion what we need is Regeneration, not Revolution. Perhaps, to be up-to-date, we may call spiritual regeneration by the term "spiritual revolution." They are really beyond me! I would like to have some definite idea, and you must tell me, otherwise our friendship will end here to-night."

He replied, "Certainly, you know what is meant by service to the people. I am sure you can do it. If you cannot, nobody else can. Oh, don't worry! If you don't know now, you will know when you get the assignment."

In a surprised tone, I immediately asked, "The assignment? What assignment? Can't I serve the people when I feel like it? Do I have to take orders to serve others? You know the whole business; you must tell me, otherwise. . . ."

Pressed harder and harder by me for an answer, he kept quiet for a few minutes, but at last said, "I should not tell you but, at the same time, I cannot be ungrateful to you for what you have done for me. I am going to let you know but you must keep my name very confidential. Your first assignment will be the indoctrination of all the Church workers and Christians in Kunming. This should be easy for you. If you prove yourself faithful and loyal, you will be made the head of the United Church in this area. If you like, you will be asked to take charge of all the religions. Then you will be great, the great leader of the religious bodies, given power to set the churches in order and clean up the reactionary elements thereof!"

What a great challenge! What a grave responsibility! While he was telling me, my mind was full of questions and doubts, and could not begin to answer any one of them. "Can I take on the work of indoctrinating Church workers and Christians? Is it one of my duties? God forgive me for helping our fellow prisoners to understand New Democracy and Communism. 'If I prove myself faithful and loyal'—does that mean that I have to fall down and worship another god? Can I serve two masters? Such a tempting offer! Position as the head of the United Church! Authority to set the churches in order! Power to clean up the reactionary elements! After that . . . what?

"Can we Christians make a compromise with New Democracy and Communism? Can I pull the two extremes or opposites together and make a new 'doctrine of the mean' which we might call 'Marxianity?' Will I be allowed by the State to do it? Should religion be subject to the State or be kept separate and independent? If I play ball with the Communists, can I be of any help to the Christian church?"

These thoughts were like tangled silk and I did not know where to begin. I was perplexed and bewildered, and did not know what to say. After a moment of silence and a deep sigh, at last I said, "Friend, I am very grateful to you for telling me my assignment and also grateful for this great opportunity. I guess I am the monkey—as we Chinese would say, 'In the mountain where there is no tiger, the monkey becomes the king.' If I were a tiger here in Kunming, there would be no question at all. What I am afraid of is that it is too big a responsibility for me—only a 'monkey'—who may spoil your nice idea and plan. Certainly, it is a great issue in my life and I must have time—plenty of time—to consider it carefully."

21

My Release and Bewilderment

ON MARCH 17, 1950, when we were ready for our first meal at 10:30 in the morning, it was my turn to clean up, with rice paper, the rice bowls for our cell mates. I finished cleaning them all quickly and perfectly except my own, which was very dirty as somebody else had used it without washing it afterwards. There was no water so I had to use a bit more force to wipe it clean with two pieces of grass paper and, by so doing, I broke it. "Congratulations, Old Bishop," a cell mate shouted.

I felt rather bad for I had no rice bowl in which to eat my meal. Turning around to him, I said, "You ought not be happy because I broke my rice bowl. Instead, you should be sorry for me."

"Oh, no, Old Bishop," he exclaimed. "To break your rice bowl is a very good omen. According to the prevalent superstition in jail, when a prisoner breaks his own rice bowl or teacup, his days of confinement come to an end. It means that he is soon to be released."

"I wish it could be true," I replied, laughingly, "but, sorry, I don't believe in superstitions." While I was swallowing my rice with my teacup, the assistant to the first secretary in the courtyard downstairs called my name and said, "Pick up your stuff and come outside immediately, for Minister Liu's sedan car is waiting for you outside."

"I don't know Minister Liu," I yelled back. "Why should he send his sedan car for me? Please don't make fun of me!"

Rushing into the jail, the first secretary also shouted, "Hurry up, Old Bishop, take your belongings and come out. The chauffeur cannot wait too long for you."

Half believing and half doubting, I took only a blanket and a copy of the New Testament. I was so dizzy that I practically rolled down the staircase, and slowly walked to the prison gate. True and sure, there was a Buick sedan waiting for me. When the chauffeur opened the door and told me to get in, I hesitated because my clothes and blanket were so dirty. Following behind me, the First Secretary practically pushed me into the back seat, and he got into the front seat with the chauffeur.

Soon the motor started and we drove from one street to another. What it was all about, I did not know and had no way to find out. At least I had one consolation—it was a sedan car and through the windows I could see whither we were driving. On the main street, it was different this time. Two members of our Church saw me in the car and looked very much surprised, and anxious to speak to me, but the car glided by and there was no chance. After a few more small streets, we arrived at the gate of the army jail. I was more puzzled and my fear became greater, as I had heard that the army jail was one of the worst jails in Kunming during those days of terror.

Getting out of the car, I was told to take my things and rest in the open courtyard inside the main entrance, where the second boss, Deputy Minister Jen, was lecturing to a group of twenty or so prisoners. I could not stand very long!

Without knowing what to do or what the outcome was going to be, I laid my blanket down by the side of a wall and sat down on it to wait, wondering what they would do with me.

A few minutes later I was asked to give two or three names of the persons living in the Diocesan House to the chauffeur so that he might drive over and get them. In the open courtyard, sitting against the prison wall, I was wondering in my mind, "Are they going to get those persons from the Diocesan House to bear witness against me at the forthcoming trial? Am I transferred here to be indoctrinated some more and with more forced labor? If they intend to release me, why should they drive me here to the army jail? Why did they bring me over in Minister Liu's sedan?"

I waited and wondered and could not make out the "dose" they were giving me. No one spoke to me or paid any attention to me. Even Mr. Ho, our theological student, working in his office about ten yards away, looked at me sitting against the wall and immediately turned his head away, pretending not to have seen me.

Forty minutes later, two representatives from the Diocesan House arrived and we three were ordered to see Minister Liu, but to listen and not talk. Following the guide (neither a soldier nor policeman this time), we entered a very small room, about ten feet square, in a separate building in the middle of the open courtyard. It was decorated in the simplest way, with only unfinished furniture including one desk, two chairs, one bench, and one wooden bed made of three or four pieces of boards on two benches. That was both his office and bedroom.

Seeing the three of us, the Minister slowly got up and politely asked us to sit down. As I was very dirty, I asked to remain standing but he insisted that I should sit down, too.

Then he started pacing back and forth in the small room and making a few remarks, saying, "Your case is cleared now. Your trouble is with the Christians of your own Church. The new government is always just; it is the government of the people. You must uphold the Common Principles and you must serve the people. I am going to release you and give you time to consider. Now you may go back."

He opened the door and ushered us out. Thus I was released without a chance of finding out how I could serve the people, and how much time I would be allowed for my consideration! We had been told to listen only, not to talk.

The news of my release spread like wildfire. Having heard that I was declared innocent, released, and had come out in Minister Liu's Buick sedan, my friends and fellow workers and Church members were no more afraid of me and very anxious to know the story. Consequently, my house was swarming with guests, among whom was a physician of our Church hospital. He gave me a simple medical check-up and found that I had lost 27 pounds, suffered a high-tone deafness, had shingles around the left side of my waist, and a big flesh hump on my back. He concluded that it was all due to malnutrition and nervous tension, and issued a medical order that I should go to a country place and have good food and complete rest for at least one month.

The recovery of my health did not seem as important as the arrival at a definite decision whether I should accept such a tempting offer or not. This worried me greatly. However, with that medical order, I could expect to have at least one month to consider the proposition.

Fortunately, too, that evening, one of my Communist friends working at the Ministry of Public Safety came and called on me, for what purpose I did not know. Taking the opportunity, I showed him the medical order for complete rest for one month. Immediately he approved the order and gave me his authoritative consent by saying, "You are in bad shape. You should have one month's complete rest. I shall report it to the Ministry of Public Safety and I am sure it will be all right." Before he went away he said, "Have a good rest. I'll see you one month later."

Then I was indeed at the crossroads! I did not know what to do. I could share with no one my inner problem of accepting or rejecting such a tempting offer. My friends and colleagues believed I needed merely physical recuperation, that in time I should be back again to take up my responsibilities. For the time being, the members of the Executive Committee kindly consented to remain functioning in my absence. But they were all interested in my life in jail and requested me again and again to tell them about it.

Knowing very well that the Communists would like to know my reactions, my mouth was closed and I revealed nothing, although I told my clergy that it was the best training camp for the servants of God, where I had had plenty of time for meditation and prayer, which brought me much closer to God, and where I had plenty of opportunities to undergo as well as to see human sufferings. Only through actual sufferings can we develop our sympathy—not pity—for the suffering and find the real meaning of the sufferings of life. "Whom the Lord loveth, He chasteneth." It was the place where you would find more chances to serve the sick and console the mentally wounded than you could handle, and where your faith in God would either be much strengthened or totally broken.

In other words, you could not remain lukewarm. That was where I experienced the imperative value of faith, without which people almost daily attempted to give up their lives, preferring death to life. Only with faith in God, I pulled "through the valley of the shadow of death."

It was also the place where I learned to appreciate freedom. Many a time I wished I could have two wings on my back like a little bird, flying freely in the sky and looking down on the stupidity and brutality of human beings lusting for power and worldly possessions. In short, I told them, "I don't regret my imprisonment and am very happy and grateful to God for giving me that opportunity. To learn those valuable lessons which no money can buy, I feel that every servant of God should go and stay in jail for a couple of weeks!"

While I was talking with my friends and colleagues, and seeming very happy, in the depths of my heart I was constantly bothered about this tempting offer. It sometimes thrilled me when I thought of the power, prominence, and ability I would have to help the churches and, at other times, it caused doubt, suspicion, and fear in me that I might be trapped. I was very much bewildered—to accept or to reject? I wished I could see clearly the will of God. When I was left alone, I spent most of the time on my knees searching for the will of God by means of prayer and reading of His words. In prayer I learned to listen to His voice and I talked to Him with many words, beseeching Him to let me know what to do. Prayers gave me no revelation. Chapter after chapter of the Bible was read, but reading shed no light. Indeed, I was at the crossroads!

After much thinking, praying, and reading of God's words, I found out one definite thing, *i.e.,* I should never seek for power, prominence, or high position. In the sight of God, the highest should be the least. Going over in my mind all the clergy and the ministers of other denominations, I could find human weakness and faults in them, but nothing reactionary. How could I prove myself faithful and loyal if I could not even find, to say nothing of cleaning up, all reactionary elements in the churches? It was only then that I could say, as our Lord had said, "Get thee hence, Satan, for it is written: thou shalt worship the Lord thy God, and Him only shalt thou serve."

After my decision to reject such a tempting offer, I was left with two other alternatives. One was for me to remain in Kunming and get ready to go back to jail again, and the other was for me to get out and wait for the storm to settle down. I discussed this inner problem with three or four friends whom I could trust absolutely and they all agreed that the alternative was between accepting the offer and getting out because it would be not only more dangerous for me but also very dangerous for those who were closely associated with me, particularly those whom the diocese had sent to the States for further studies or observations, if I rejected the offer and remained. One of them even pleaded with me to go, saying, "If you definitely decide that you cannot play ball with the Communists, we promise, before God, that we'll carry on the work faithfully during your absence."

It was very easy to say "get out," but hard even to conceive the possibility. From Kunming there was no plane, and only three highways: east to Kweichow, Kwangsi and Kwangtung; south to Indo-China, and west to Burma by the Burma Road. People had tried the easten route but they all returned, for there were too many robbers, bandits, and disbanded soldiers. One of the prevalent stories was about a fellow whom we knew, a native of Kweichow. After the Communists had taken over both Kweichow and Yunnan, he decided to join his family in Kweiyang and started on the journey all by himself, saying to his friends confidentially, "Why should I be afraid of robbers and bandits? I won't bring anything with me except enough for traveling expenses!" Before he had completed the distance of 160 kilos to Tsan-Yi, he was robbed eight times and everything he had on him was taken away except a pair of short pants.

On the southern route to Indo-China, battles were still raging between the Communist and Nationalist forces. Those who had tried this route were taken and imprisoned by either one side or the other as spies of the opposite.

The western route—the famous Burma Road to Burma—was about 700 miles long. The Communist occupation of the towns along this highway was complete, with heavy garrisons both in Paoshan and Hwanting (border towns between China and Burma), and with groups of soldiers patrolling the line both day and night. Furthermore, there was no means of transportation, for the new government had commandeered all private and commercial trucks.

After I had been told of the condition of the three highways from Kunming, my hopes of getting out went to the winds. Sometimes I joked with my friends, saying, "Unless God is going to send me a pair of wings and teach me how to use them, we are all stuck here for better or for worse. In other words, as we Chinese would say, 'Our fate is sealed.'"

22

House Arrest and Welcome Party

ON SUNDAY morning, March 19, just previous to the ordination service at our Cathedral advancing one deacon to the priesthood and ordaining two to the diaconate, one of our colleagues told me about a couple who were staying in the meeting room behind the Cathedral opposite the Deanery. In pride and pleasure, he said, "The girl is 'so and so' (I don't remember her name) who has been singing in our Cathedral choir, and the boy is one of the children of the postmaster at the Ching Pi Road branch post office. They were married not many days ago without any ceremony. Oh, they are so friendly and have been so helpful to us!"

When I heard that they had been married without ceremony, I began to suspect them even before I actually met the boy at the service. (Note: The members of the Communist Party may get married without any ceremony provided they have secured approval from the Party, followed by an announcement to the members of their cell. Wives are called lovers only.) Another secret I learned in jail was that the more friendly the Communists were toward you, the more probable it was that they had been assigned to spy on you. Sooner or later they would fulfill their mission and you would become their victim and, thus, they would get their credits or merits. You could not get away from them! In dealing with them you had to watch every word you spoke and every action. Indirectly, I warned my colleague, "Watch what you say and do very carefully."

Sure enough, that night when I was asleep, I heard somebody pounding on the main gate of the Diocesan House, which woke up everyone in the compound. As soon as a light was turned on (it was two o'clock in the morning), I saw our colleague, pale and trembling, who had come over with his Communist friend to get my advice. According to his story, that night about 11 o'clock our Cathedral was surrounded by soldiers and in came six fully armed youngsters who searched both the Cathedral and the Deanery, on the ground that our colleague's name had been called on the air by some secret radio station. "Although they found nothing," my colleague said in a trembling voice, "they planned to take me away. It was very fortunate that my friend [pointing to his Communist pal] came out and guaranteed that I would not run away, otherwise I would be in jail now. What should I do, Bishop?"

I replied, "I cannot do a thing for you except pray, which I'll do. My advice is, 'tell them all you know and tell them frankly.' Have a clear conscience before God, don't be afraid, and you will be all right!" After a moment of thinking I commented, "It seems strange to me that if you have some connection with the secret radio station, they should call your name in Chinese instead of some number. Have they heard your response on the air, too?"

"No," he answered, "they did not say they had heard."

"Go back and rest. Hold on to God and don't be afraid. You have a good friend—I am sure he can help you," I said.

Standing by and listening to our conversation, that Communist friend of his appeared very much displeased with my comment about calling his full name instead of some number, etc. He turned to me and said, "You stay here and don't go out, until I have cleared the case." Turning to my colleague, he said, "Let us go back home and then let me go to the Ministry of Public Safety and see what I can do for you." So they left and thus I was under house arrest!

My colleague was very happy to be able to come back to see me on Monday afternoon, March 20, telling me that his case had been all cleared, and that I would be free to go out. "Oh," he said, "I am so grateful to him for putting in a formal guarantee for me. In return, I had only to promise to do everything he would like me to do. Tonight I shall go to the broadcasting station to give a ten-minute radio talk to the Christians in Kunming on the privileges and benefits of purchasing the Victory Bonds."

I was glad to hear that I was freed from house arrest but sorry to learn that he had promised to do everything he would be asked. Realizing that he would be used by the Communists for propaganda purposes, and possibly for creating trouble within the Church, I said just a few words to him, "Watch out for yourself! Remember, you are a servant of God."

That afternoon another guest came in to see me, Dr. Chi, a physician of our Hueitien Hospital, for whom I had done all I could to help when he was stranded in Kunming, having no business at all with his private practice. He told me, "The Hospital Union is planning to have a big welcome party in your honor on Friday night. You had better be prepared."

Hearing about the welcome party, I was really sore within myself because of the subtle methods by which the Communists tried to destroy human dignity and self-respect, and said, "Welcome party in my honor! Such a beautiful word! Where is the 'honor'? I am ashamed of my imprisonment. There is no honor about that! Why don't you call it, 'Humiliation Party' instead? As a friend for whom I have 'gone all out' to help, you ought to tell the truth and tell me the facts. You said, 'You had better get prepared.' From that statement, it seems that you belong to the inner circle. If you are a friend, tell me the truth; if not, don't tell me anything."

Pressed hard by me, he began to confess that he was a member of the Youth Corps, together with a number of doctors and nurses at the hospital. To express his gratitude, he continued, "Because I am grateful to you for what you have done for me, I am here telling you to get yourself prepared for that 'humiliation party,' as you have named it."

"Don't worry. I know what I am going to say. I heard of this so-called 'welcome party' in prison. You don't have to humiliate me and you can't do it. If you members of the Youth Corps want to take the hospital, you may do so. I would be glad to get rid of this headache. I will say what I have said in my confession paper, and no more. In my confession paper I did say that sometimes I was a little too strict with my fellow workers. I'll say the same thing, that is all. I guess you members of the Youth Corps have read my confession paper, too."

"Yes," he said, "we have read your confession paper. This is just what we want you to do—to give up the hospital to the government and to confess that you have been a little strict with your subordinates. That is fine!"

"Look, Doc," showing him my "stick legs," my shingles on my left waist, and the flesh hump on my back, I said, "my health is a mess. I have a medical order to have one month's complete rest and also approval from the Ministry of Public

Safety. I must have it before I do anything at all. Please ask them to postpone the 'humiliation party' until I come back from my recuperation." After examining my physical condition, he agreed and consented to do his best to get it postponed.

Purposely I changed the subject of our conversation and said, "Doc, I know that you are a very devout Christian and now are an associate member of the Communist Party. Do you still believe in God and the future life, and that you are a child of God?"

"Yes, I do," he answered, without the least hesitation.

"Then how can you reconcile the two—Materialism and Spiritualism, Communism and Christianity," I asked again. "To me they are absolute opposites. In prison I was very much amused by a Communist who was afraid of ghosts. For at least the past two months I made a very careful study by comparing the differences and similarities between Communism and Christianity. I am still unable to make any reconciliation between the two. Oh, I am so dumb! I cannot discover a new theology which we may call 'Marxianity.' As a devout Christian and a faithful member of the Youth Corps, you must have the secret of making such a compromise. Will you give me some light?"

He replied, as most Chinese would reply, "I am not a theologian, nor do I think much. I believe in what is good for me; that is why I believe in God. But the present current is such that there is no use for us to row our boat against it. Be practical, Bishop! We have to survive!"

"Have you read the last page of the book, *New Democracy,* by Mao Tze-tung?" I asked. "There it is clearly stated that no member of the Party will be allowed to have anything to do with religion. Suppose, in the future, the new regime would require every member of the Party to sever his connections with religion. What would you do?"

Very much embarrassed, he started to scratch his head a little and, at long last, jokingly said, "We just wait and see. At least one thing I am sure of is that they don't have a microscope to examine my heart to see whether I have any religion there or not!"

I politely explained, "Friend, I have no intention to embarrass you. What I want is your secret, if you have any, to help me solve my inner problem. Frankly speaking, the more I have thought of this question, the more I have come to the conclusion: 'No man can serve two masters.' 'He who is not with Me is against Me.' It is true with Christianity and so it is with Communism!"

23

God's Will Finally Revealed

TO AVOID further complications and the wear and tear of talking, and to have some real rest and recuperation, I had to get away. Above all these things, I needed a quiet time to search out the will of God for the decision which I had to make before the end of one month. So I decided to accept the invitation extended to me by an old friend of mine who had a house in the suburbs. Because of our differences in Church affiliations and professional interests, this friend and I had for years seldom crossed each other's paths. Somehow, he heard of my release and the necessity for recuperation and sent me a message on Saturday, one day after I had come home, saying, "My home is ready to welcome you; one thing we are sure to provide is rest and quietness for you." God was certainly gracious unto me and did provide every need of mine.

On Tuesday morning, March 21, at 7:00 o'clock, when the people, old and young, in the Diocesan Center were still busy in their apartments and houses with washing and cooking breakfast, and the sun was just out, I took a small suitcase (which I had packed the night before with two pairs of socks, two suits of underwear, two shirts and one copy of the New Testament in Chinese), and walked quietly through the back door out into the small alley. Treading along the gravel-paved highway, I soon arrived at my friend's house and was warmly welcomed and immediately put into separate quarters prepared for me. Except for three meals a day, when I joined the family, I was left absolutely alone and cut off from the world. I had the best food of my life, with many northern, southern, and Yunnan dishes spread on the table at every meal. My appetite was good and I ate like a pig.

Save for the members of this family, nobody knew where I was. With no visitors to see, no headaches and problems of the Church to be solved, no mental torture to go through, no threats to bear, and no sufferings to see and hear, I was really in paradise. My daily schedule was eating, sleeping, praying, and reading the Bible. Whenever I was awake, I was on my knees praying and reading the words of God. I prayed and prayed. In my prayers I cried to God beseeching Him to reveal His will to me, and I learned to listen but no voice was heard. I hoped anxiously to get some direct light from my reading.

The Psalms became my favorite and I loved such passages as, "God is our hope and strength, a very present help in trouble; therefore will we not fear, though the earth be moved and though the hills be carried into the midst of the sea"; "He truly is my strength and my salvation; He is my defense, so that I shall not fall"; "Thou art my hope, and my stronghold; my God, in Him will I trust. . . . He shall defend thee under His wings, and thou shalt be safe under His feathers; His faithfulness and truth shall be thy shield and buckler. . . . Thou shalt not be afraid of any terror by night, nor for the arrow that flieth by day. . . . A thousand shall fall beside thee, and ten thousand at thy right hand, but it shall not come nigh thee."

From my readings, though I got much consolation and

hope, and above all, *courage*—as I should have—I was not satisfied because I did not get a direct and definite answer from God as to what I ought to do. Many a time I wished I could have a direct wire with God! Many a time I was frustrated, doubting whether God was really taking care of me, or blaming myself because my spiritual ears were too undeveloped to hear His voice!

At last I became desperate and as human as any human being could be, and put my problem of making that decision squarely before God. In my own words often I prayed, "God, I fail to get Your answer to my prayers. You have told me only to have courage, and nothing else. Excuse me for passing the buck to You. If it is Your will for me to remain in Kunming, don't do a thing for me. However, if it is Your will for me to get out, please provide the means of transportation, and don't give me any trouble on the road."

With this simple and childish prayer of mine often repeated, I began to feel tremendously calm and concluded to leave the decision to God whether I should remain in or get out of Kunming. At the same time I told myself that, as God has taught me to have courage, I must have faith to coöperate with God in fulfilling His will in me. I am not going out to seek for means of transportation, but wait and see.

After a few days of such a serene life, I picked up some weight and strength. I had eaten enough, and slept enough. Occasionally, then, I visited my host and hostess and discussed my problem with them. Both of them insisted that I should get out, since I had that tempting offer from the Communists. "From what I have heard," my host said, "if they have their eyes set on you, by hook or crook they will get you. There is no way to refuse them. If you do the work for them, you have practically to denounce your faith, speak against your conscience, and probably kill many of your Christians and Church leaders. By all means, get out! But how can you get out, is the question. There are no means of transportation and the highways are full of robbers and bandits."

Amazed at the speed with which he arrived at his definite decision for me, I said, "Friend, I am really surprised that you can take this issue and reach your conclusion so easily and quickly. Look here, I am the head of our Church and shepherd of my flock. In the time of distress I should stay on more closely with them and not leave them to the wolves. If I try to escape and am caught, I would be finished without question. Furthermore, I don't know the will of God, whether I should remain or get out. That is why I have been unable to come to any definite decision yet. Besides, I don't see any possibility of getting any means of transportation."

"It is true," he replied, "that if you escape and are caught, you will be finished. However, the chance is fifty-fifty. On the other hand, if you remain in Kunming, there is no work for you to do and you cannot be shepherd to your sheep. Instead, you will be forced to be their chief enemy, giving them trials and condemning them to imprisonment or death. When your work is done, you will be not a martyr of faith but a victim of political conspiracy and a traitor to your people. I wish I could send you to the border. The trouble nowadays is that our friends who have trucks are not permitted to move unless sanctioned or ordered by the government."

My hostess, trying to comfort me, concluded, "Don't worry; there are still three weeks to go. God will provide some means, I am sure, if we pray earnestly enough. I'll see what we can do."

Sure enough, God answered our prayers, not in the ways I asked but in much better ways than I had ever dreamed or expected. On the twelfth night, a Presbyterian lady named Mrs. Lydia (fictitious name), who visited my host and hostess occasionally and had learned of my difficult problem, voluntarily came to my quarters and said, "Bishop, you must get out. There is no more work for you to do. I have a friend who has been ordered by the new government to go to Burma to-morrow with his three trucks. There are loads and loads of cotton yarns stored in both Hwanting and Chiukoo waiting for immediate transportation back to Kunming. He is a very nice man and a good friend of mine, whom you can trust. He will be on one of his three trucks. He is allowed to take passengers in order to cover expenses, as the new government subsidy is not enough. I also have many friends along the Burma Road. I shall be glad to accompany you to the border and give you all the assistance I can. Are you strong enough to travel? Can you get ready to go tomorrow morning?"

Overwhelmed with joy at her gracious offer to accompany me to the Burma border, I did not know how to express my gratitude enough to her, but just said, "You must be an angel sent by God to escort me. If I am ever caught, you can be assured that I'll *never, never* mention your name!"

"Don't worry, Bishop," she said, with a big laugh, "we shall get through. I have been on that road—I cannot remember how many times—and I never had any trouble. After we get on the truck, you just keep your mouth shut and don't reveal yourself as a bishop. Leave all the rest to me."

After thinking, and talking for a while with my host, Mrs. Lydia decided that we both should ride in a jeep, to avoid the first checking station outside the west city gate and evade the second station on the Western Hills, and go to a small village about twenty miles west of Kunming first. There we would wait for the trucks. She went back home and made the necessary arrangements.

After a short prayer, I went to bed but could not sleep, so continued offering my thanks to God for answering my prayers in such a mysterious way and beseeching Him to grant us protection on the way. I was then deeply convinced that it was the will of God for me to get out, otherwise such things would not have happened—a Presbyterian woman, whom my wife and I had known only a couple of years, willing to leave her family and risk her life in order to escort me to the Burma border! Mysterious! Miraculous! Doubts and fears were gone. With a grateful heart and peace of mind, feeling only unworthy of the grace of God, I patiently waited in bed for daybreak!

24

My Escape

PUNCTUALLY at 8:00 o'clock next morning the jeep came. After our farewell prayer together, we got in and the jeep rolled along the Ching Pi Road and then Tai Ho Street, two of the most prosperous streets outside the Kunming city wall. I found a few differences on the streets. There were more people walking back and forth in the middle of the street, so we had to use our horn constantly in order to get through; secondly, about one-fourth of the shops were closed; and last, not infrequently we saw groups of three or four men in Lenin uniforms marching proudly down the streets, with their pistols on their left sides.

Soon we finished the detour and by-passed the checking station outside the West City Gate. On the highway to the Western Hills our jeep rolled on faster and faster and we began to climb up higher and higher. On one of the "hairpin" curves facing Kunming City, I said, "Farewell, Kunming. May God bless you, and strengthen the God-loving people in the city in their struggle against the ungodly!" Before we got to the checking station on top of the hill, our jeep had slowed down considerably. We expected to be halted, but instead, the soldier on guard saluted me and let us pass without saying a word. I nodded my head as if I were a big shot closely connected with the Communist army.

At twenty minutes after nine we arrived at our destination, a very small village about twenty miles west of Kunming. We picked the best tea shop and ordered the Fukien green tea to enjoy while we waited. Thousands and thousands of young Chinese Communist soldiers marched westward in groups to Tali and Paoshan. Boys and girls in the same uniform were mixed together. Some of them were sick yet forced themselves to pull along, in order to help their groups win honors in the competition of marching.

We waited and waited, and our green tea was changed to white boiled water. Still no truck came. At noon we helped ourselves to some hard boiled "tea eggs" for our lunch, hoping our trucks would arrive at any moment. One hour went by and then another, and there was no sign of any truck.

We could not stay in the small village over night as we had been told that there were too many bandits in that area. Strangers were the targets of the bandits for their personal possessions, and of the Communist patrols looking for bandits. Furthermore, we had to find out why our trucks did not come, so when the clock struck four we decided to return. On our way back, my doubts and fears crept into my mind again and I thought it might be the will of God after all for me to remain in Kunming. We both felt gloomy and bewildered, and did not speak a word the whole way back.

When we arrived at his door, my host asked, "Have you seen Lao Shan, our servant, whom I sent to tell you that they have decided to start tomorrow morning? You must have passed each other on the way. Everything is all right now, so I have heard. The trouble was that, at first, the truck owner had so much red tape to go through and so much money to pay for the road repairing fee that he had to borrow from friends. We have loaned him the money. When he finished with the red tape, they found that one of the passengers, for some reason or other, had been arrested by the police. He is still at the police station being questioned. Whether he can go with his family tomorrow is a question. However, the other passengers, together with the truck owner, have decided to leave early tomorrow morning without him if he cannot make it. You had better rest tonight for you will have to get up one hour earlier tomorrow morning. You just be at the tea shop on the other side of the Western Hills, which is much nearer, at 8 o'clock and get on the first truck."

At the dinner table that night, I casually expressed my thoughts by saying, "Friends, don't you think that it is the will of God for me to remain in Kunming? I may be arrested by the police, too, just as that passenger!"

"Don't worry," said Mrs. Lydia, "we don't want to see you stay in Kunming. There is no danger as we are not going to the police station to be checked by them. Didn't we get over the Western Hills in the jeep? Really, the more they check, the better. As soon as the trucks get moving on the highway, everything will be all right. Don't you remember the Chinese proverbs, saying, 'Everything is hard in the beginning,' and 'Good things always have difficulties'? Trust in God; you will be saved. You must have courage to coöperate with God, otherwise you can't blame Him for the consequences!"

Very much encouraged by her statements, I answered, "You certainly have the nature of God, who loves us more than we love ourselves. You are more anxious to get me out than I am myself. It is really great and miraculous! I am very grateful to you!"

In response, she said, "Yes, I *am* anxious for you to get out.

Why? I don't know myself. I feel it is both my duty and privilege to help you. You need help!"

Promptly at seven the next morning, the jeep was ready to take off again. By the same detour, we by-passed the first checking station and were soon on the Western Hills highway again. It was windy and chilly that morning. When we reached the top of the hill, the soldiers stayed inside the station and paid no attention to us. Going down hill we went as if flying. Quickly and safely we arrived at our planned destination, a small tea shop at the foot of the hill—this time at 7:40 a.m. It was too early and the shopkeeper was just getting up. No tea for us! Mrs. Lydia sat in the jeep while I stood in the cold wind, watching. It was another good exercise of patience. We waited and waited again.

It was 11 o'clock when our first truck arrived and stopped. Quickly we were pulled on from the back. It was an open truck, with a torn canvas cover laid aside, and full of bedding rolls, trunks, boxes, suitcases, and what not. The seventeen passengers, including two children, Mrs. Lydia and myself, sat in three rows on top of the luggage, with the wind blowing hard from the front, the sun shining brightly from above, and yellow dust from the gravel-paved highway rolling up and covering us from behind. It was the first time I had experienced and understood the Chinese phrase, "yellow fish." These were people who could not buy tickets from the Highway Bureau as regular passengers. By means of paying more to the driver, they occupied the back space in the truck, as a rule, where yellow dust, rolling up constantly from the back wheels, soon covered every passenger and thing there. Before long, I was covered thickly all over with yellow dust and became a real "yellow fish." Going down hill, the truck was rolling fast, as if flying, while up hill, every male passenger was required to get off and walk. The engine had no power because it was run with alcohol instead of gasoline, which had been confiscated by the new government. Slowly we rolled on, bumping up and down whenever passing over holes or stones in the road.

In the dark we finally arrived at Chu-hsiung, the destination of our first day. Under the dim light of a vegetable oil lamp, the shopkeeper cooked our meal which we swallowed down fast, as well as the dust in our mouths. We were both hungry and deadly tired, as we had been well shaken up on the road. Quickly we crawled into our beds and were fast asleep. We were indeed "dead to the world," and the world was likewise dead to us! Somehow, even the Communist night patrols did not bother us that night. But when we got up the next morning we heard the three truck drivers complaining that they did not have enough sleep, due to the fact that the Communist night patrols had inspected every piece of luggage on the trucks and interrogated the three of them for a couple of hours. I said to Mrs. Lydia, "God does protect us. It was a miracle of God that they missed us. If they had questioned me as they did one of the truck drivers, what would be the outcome?"

Next day we rolled on very slowly and the three trucks were traveling not very far apart. Thus, more dust and sand were rolled up by the wheels, but in dust and sunshine I slept most of the way, except for occasional bumps waking me up and when we were told to get off before climbing up mountains. Just before dark we arrived in Hsia Kwan, a highway junction, heavily guarded by Communist soldiers. I was still dozing and slow in getting off the truck. The Communist inspectors got on from the front for inspection

and by the time I jumped down behind, one of the policemen saw me and questioned me, "Who are you?"

Politely and frankly I said, "I am the Bishop of the Holy
Catholic Church in Yunkwei, and a passenger going to the
western part of Yunnan."

"Well . . ." he said, looking at me. "All right." This was
the first and only time I was ever questioned. He did not ask
me any more questions, although other passengers were
questioned again and again and their papers had to be carefully examined. It was only a miracle of God that I was so
well protected!

Arriving at Hsia Kwan, I was full of happy memories and
varied reflections. It was a beautiful spot, with a big lake on
one side covering several districts and a high mountain on
the other covered with snow all year round. It is one of the
very few places in China, if not in the world, where water
flows from a lake into the mountains. It was the town where
the leaders of society were once so much interested in the
work of the Church that they donated land, buildings, and
some funds for us to establish a school, a hospital, and a
church. A few miles away to the north was Tali, where we
had a church, a school, and an orphanage. To the west, just
a mile or so, there was the "Heavenly Bridge" where Kung
Ming, the greatest statesman and scientist during the period
of Three Kingdoms centuries ago, on his campaign against
the revolts of the tribal people of the Southwest, captured
and released Meng Ho, their rebel leader, seven times.

Pointing to the "Heavenly Bridge," I said to the truck
owner, "That is the place where the leaders of the new
regime should come and compare their policy of governing
the people by slaughter and force with that of Kung Ming
who, instead of killing Meng Ho, as he could have done

easily after the first capture, released him seven times in order to win his heart. As Kung Ming said, 'I may kill one Meng Ho or two, but many more Meng Hos will emerge and lead their people, if I don't win their hearts.' "

That night we all went to the biggest restaurant, where we cleaned ourselves up thoroughly and ordered the best food on the whole trip, including the famous dish, "Hsia Kwan"—fish from the lake boiled together with soy bean curd. While we were enjoying the delicious food, our truck driver came over to our table and asked if any one of us would care to stay on the truck that night so that he might go back and join his family in Hsia Kwan. Happily, I volunteered to sleep on the truck and take care of it and all its contents for him. He was so grateful to me for being able to join his family that night that we two became very friendly, and later he did all things possible for me—help, as well as comfort.

Therefore, that night I slept on the truck very comfortably and in peace, without any interference, while the other passengers staying at the hotel, including Mrs. Lydia, were questioned one by one by the Communist night patrols. With her incomplete papers, one lady passenger would have been arrested if it had not been for her crying and weeping, which aroused some pity or sympathy on the part of the patrol leader. I thought afterwards of myself—if I were caught and questioned as that woman passenger, when I had no papers whatsoever, what would I do? Instead of receiving thanks from the truck driver, I should have given him mine. How mysteriously God worked to protect His unworthy servant! I was saturated with gratitude and praises to the Lord!

After Hsia Kwan, the Burma Road started getting worse

and narrower and our speed had to be cut down even more. The destination of our third day was Paoshan, the most important center of Western Yunnan, where the Communists had stationed a huge garrison force for the purpose of controlling Western Yunnan and the frontier between Burma and China.

For some reason or other—I still don't know what—Mrs. Lydia insisted that I should go with the group and stay in the hotel over night. This was the second and last night I passed with the group at a hotel where two night patrol groups came, one military and the other the city police. The former was in too much of a hurry while the latter was careful in going from room to room, but somehow neglected two rooms that night. That I was in one of the two rooms missed intentionally or unintentionally I had no way to find out. Again by the grace of God, I passed the Paoshan "Pass" successfully.

After we got through Paoshan, all was smooth and quiet until we reached the border town of Hwanting. Communist soldiers were marching back and forth on the highway and patrols were constantly seen in trucks and jeeps. They did not bother us. Although we spent three more nights on the Burma Road, no one was checked or questioned. The farther south we went, the hotter it became. For the next three nights I slept on the truck with the driver and enjoyed the cool breezes and the bright stars above. Only one night (the fifth) were we disturbed and frightened, including the people in the village, by a skirmish between the Communist soldiers and the tribal people under the leadership of their Tu Shih (head of the tribal people, who had power of life and death), who were trying to get some ammunition away from the Communist soldiers.

The highway was in poor condition, as if it had not been repaired, at least for months if not for years. Some bridges were worn and others gone entirely. Many a time our truck had to drive through the shallow water of creeks. One great difference before and after the Communists took over the Yunnan Province was the growth of poppies along the highway. Before the establishment of the Communist regime, I had traveled back and forth on the highways many times and had heard of plantations of poppies, but they were in the most remote areas of Yunnan where practically no people would go—never along the highways where travelers could see them easily. On my escape along the Burma Road, we saw nothing but field after field of poppies, blooming beautifully. When the wind was blowing gently, they were but waves of "poppy seas." (Poppies are used to make opium.)

On the morning of the seventh day, we got up late as there were not many miles to go to the border town called Hwanting. On one hand, we were all happy for covering almost 700 miles on this famous (but gravel-paved) Burma Road without a break or any engine trouble, for meeting neither bandits nor robbers, for getting through so many Communist checking stations and passes without any unhappy incident, and above all, for the hope of getting away soon from the Red grip and breathing the air of freedom again in Burma. On the other hand, we had heard that the border town of Hwanting was very heavily guarded by Communist policemen, guards, soldiers, secret service men, and what not. It was the last and hardest "pass" I had to get through.

On the truck I worried, feared, and my heart was pounding in my throat. Constantly I prayed in my mind, beseeching God to help me get through. Also, I was repeatedly encouraged and inspired by Mrs. Lydia's faith and words,

"Don't worry; you will get through, for God is with you."
At the same time, she whispered the directions to me as to
how to get through, which I memorized very carefully.

When Hwanting was in sight, I started getting myself
ready, first, by cleaning up a bit with my handkerchief. Then
I picked up the jar of vaseline in which I had hidden my
episcopal ring, and put it into my left trouser pocket. At a
distance of about ten yards from the checking station, the
truck stopped and I jumped down from the back. It was two
o'clock in the afternoon and so hot and humid that all the
soldiers, policemen, guards, and even the staff members
working at the Custom House, were taking a nice siesta,
except one policeman guarding the China end of the iron
bridge over the creek dividing China and Burma. He was
talking to a beautiful young girl and paid no attention to me,
so I walked first through a short street and then over the
international bridge without any trouble!

As soon as I landed in Burma, I jumped up, kicked my
legs, stretched my arms, took a deep breath, and said a short
prayer, "Thank Thee, O God, for Thy protection and my
freedom restored, which I'll enjoy, appreciate, and treasure
always!"

In spite of the fact that by a series of miracles of God I had
broken the Bamboo Curtain and thrown away the Red chains
from my legs and mind, my troubles did not end there as I
did not have my passport with me or a visa from the Burma
government, or any kind of paper of identification. As a
stranger in a strange land—and a small village called Chiu
Koo, where there was neither any postal service nor telegraph
office by which I might send "SOS" messages to our friends
in Rangoon or Hongkong—I could be easily and legally
imprisoned in Burma. However, after my escape from the

strong Communist grip, my faith in God was much greater than ever before. Slowly and happily I climbed up the hill and there I saw a few rows of simple houses at a distance of thirty or forty yards away, looking just like streets in the villages of China. I did not mind the hot sun beating down on me or the fact that I was perspiring heavily. Step by step I soon arrived at Chiu Koo and picked a big, cool coffee shop, where I sat down with my coat off and began to enjoy myself.

A few minutes later, a young chap about thirty years old, whom I did not know at all, came walking down the street and suddenly stopped and looked at me again and again. At last he walked over to the coffee shop and asked me, "How are you, Bishop? What are you doing here?"

I got up, but was dumfounded and did not know what to say, because I did not want to reveal who I was. Furthermore, after having been some time under the Communist regime, I did not like to answer those questions when asked by a stranger. So I pretended not to have understood his questions and asked him, "What? What do you say?"

"You are Bishop Huang, aren't you? Surely, I know you because about a year ago I was in Kunming and attended Sunday services regularly at St. John's Cathedral. I am Chang Yi-hsin, a member of the Holy Catholic Church in China. I am also a graduate of St. John's University, Shanghai. My father is 'so and so' who has been serving the Church for years, and is still very active in the Church." Realizing that he was the son of Mr. 'So and so', one of the pillars of our Church in China, I began to have some confidence in him. After all, we were both in Burma, so I answered his questions confidentially and told him about my problem.

Confidently he replied, "You don't have to worry here. I

can render this little service to you easily. I have been here several years, going back and forth between China and Burma, and know practically everybody here. Give me one hundred rupee (Burmese currency); I'll have everything fixed for you."

I grasped the opportunity and also requested him to help another passenger, and he left with the money. Only four hours later he came back, not only with a permit for me to travel freely in Northern Burma for three months, but also one for my friend.

To me, it was no less than another miracle of God. It was absolutely true that by faith in Him I lived and moved and had my being!

Only by faith, I was able to pull through physical sufferings and mental tortures of those seventy-nine days in jail; by faith, I was enabled to see and resist such a tempting offer from the dialectical and diabolical "devil"; by faith, I had courage to escape when God provided the means of transportation; by faith, I came to this great land of liberty and democracy with my family, and, above all, by faith in God, I came to realize that it was my duty to tell, through my own experiences and insight, what Applied Communism actually is, in contrast to the sweet "theoretical" Communism, with fond hopes and fervent prayers that you, as the readers of this book, will not be fooled or deceived by the clever Communist propaganda and sweet promises, and so that you may learn the lessons of life, but not in the hard way—as I have learned them, at the hands of the Devil! By God I am spared! To God I dedicate my life again!

Part II

COMMUNIST LAND REFORMS IN CHINA: SUBTLE PROCEDURE AND REAL OBJECTIVES

COMMUNIST LAND REFORMS IN CHINA

Subtle Procedure and Real Objectives

CHINA, as is usually claimed, has a history of 5,000 years. Ever since the reign of Emperor Sheng Nung, "the Divine Farmer" and inventor of agricultural implements (to be on the safe side—in the 28th century B.C.), China has been an agricultural nation. Even today more than 80% of the population are farmers, solely dependent upon land. The importance of land has been a factor so closely interwoven with the national life that almost every dynasty or revolution, whether successful or not, has advocated or carried out some kind of land reform. In the nation's long history there have been at least a dozen land reforms before the "Principles of Livelihood" of the Nationalist Party and the "Tu Kai," or land reforms of the Communist Party in China today.

Let me just give you the names of the important recorded land reforms during the past 5,000 years and you will readily see that the land reforms first advocated by the Communist Party in China were not strange or unwelcome to the majority of the Chinese people. Before the third millennium B.C., lands were managed on the principle of "public ownership and public tillage." Later, in order to encourage agricultural production, this was changed to "public ownership and private harvesting," with a small tax to the government.

Then it was followed chronologically by the following main land reforms:

> "The 9-Square Land System" of the Hsia (2205-1766 B. C.), the Shang (1766-1122 B. C.), and the Chou (1122-249 B. C.) dynasties.
>
> "Taxation on Mow-unit" of Hsien Kung of the State of Lu.
>
> "The Unlimited Private Ownership and Development of Land" of Shang Yang of the Ts'in dynasty (249-210 B. C.)
>
> "The Emperor's Land System" of Wong Mang (9-23 A. D.)
>
> "Land Distribution and Ownership by Lot-Casting" of Tsing Wu Ti (236-290 A. D.)
>
> "The Equalization System of Land" of Northern Wei in the 5th century.
>
> "Distribution of Lands According to Mouths" of the T'ang dynasty (618-906 A. D.)
>
> "The Public and Peoples' Land System" of the Sung dynasty (960-1279 A. D.)
>
> "The Square Land System" of Wang An Shih (1020-1089 A. D.)
>
> "The Land Limitation System" of Chao Tien Ling, under Emperor Shih Tsu of the Yuan dynasty (1215-1296 A. D.).
>
> "The Land System of the Heavenly Dynasty" proclaimed by Hung Hsiu Chuan during the time of the Tai Ping Rebellion (1848-1864 A. D.), which advocated "land for all to till, food for all to eat, clothes for all to wear, and money for all to spend. It is to make no place unequal and no man hungry or cold."

With this long list of land reforms in the history of China, the people, especially the intelligentsia, did not feel anything strange or novel in the program of the Communist land reform. Rather, it was nationally recognized as a necessity in order to elevate the living standards of the poor, since Mao Tse-tung and his colleagues had repeatedly emphasized that the purpose of land reforms was to free agricultural production from feudal bondage. Furthermore, since 1927, the Communists have carried out land reforms in those areas under their control from time to time. With more than twenty years' experience in this work to their credit, they should have known better than anyone else how to carry out land reforms successfully when they took over the whole country, so as to increase the agricultural production and improve the living conditions of the downtrodden. Although not much was known of the actual happenings, that was the general opinion and conclusion of the educated. In short, land reforms were favored; many of the rich were ready and happy to give up their possessions. Then, in many Chinese minds as well as in the minds of so many well-wishers of China and of the Communist Party, the Communists in China were once considered only as agrarian reformers and not a part of international Communism.

Being considered by my fellow prisoners as one of the few authorities on New Democracy and Communism during the last days of my confinement, I was fortunately permitted, with a group of educated prisoners, to join the advanced group in study and discussion, and sometimes in listening to lectures on the Communist land reforms. I was shocked as well as thrilled several times to get indirectly from those study meetings a few hints or secrets as to how to carry out land reforms.

Again, I was blessed by the fellowship and friendship of one of my fellow prisoners, who had been a Communist himself and served as confidential assistant to a Communist director in the work of organizing and training the members of the Farmers Unions and carrying out land reforms in a number of districts in the Northeast. According to his story, after the completion of his assignment in the Northeast, he was ordered to Peiping to wait for new assignments. During his short stay in Peiping, he became so disgusted with the Communist brutalities—as he said, "My conscience hurts me all the time"—that he secretly left the Red national capital and fled to Kunming, trying to regain his inner peace. Unfortunately, one year later Kunming was put under the iron grip, and he was caught and imprisoned. Once in a while, he revealed a secret or two, and always commented as his conclusion, "to do such work, you have to darken your conscience!"

With the hints learned from that study group and the secrets from this fellow prisoner, which I compared with utterances and confessions of the leading Communists on land reforms and checked with the reports of many "scapegoats," friends and relatives of our friends, and with letters about a few victims sacrificed on the altar of land reforms whom I knew well, I gained the necessary insight and am able to "expose the Fox's tail" by presenting to the readers the subtle procedure and real objectives of the Communist Land Reforms in China.

After the capture of a city, town, or village by the Communists, three things are immediately taken up simultaneously, i.e., putting up wooden boxes in various sections of the city, organizing various kinds of unions, and taking away arms and ammunition from the people.

The wooden boxes (the number of which is dependent upon the size of the city, town, or village) are the so-called secret-reporting boxes. All the people are told, and many are required—particularly the former civil servants of the Nationalist government—to report by means of these secret reporting boxes all the reactionaries, imperialists, feudalists, landlords, leaders of the society, those who are or are suspected of being detrimental to the New Regime, and even those with personal grievances. "Just drop into one of the boxes the names and addresses, with charges whether true or false. You may or may not sign your name or address for future reference. If you do sign, a person so reported is counted a small merit to your credit." A policeman of the lowest rank of the former government is required to report twenty such persons in order to redeem his sinful past. Pretty soon these wooden boxes are full of the names of the "enemies of the people."

After having collected the names, the Communists, organized in different groups, go around from house to house between midnight and five o'clock every morning and arrest the people and put them all into jails, as one of the Communist principles emphasizes, "It is better to imprison 10,000 innocent than to let one guilty escape." Consequently, the prisons are packed with prisoners like sardines, and new prisons are established in schools or institutions. This also fulfills the Communist definition of "State" as consisting of only three classes of human beings: soldiers, policemen, and prisoners. The people are either soldiers fighting against the enemy or policemen constantly watching and reporting others, while the reactionaries constitute the prisoners locked up. Thus the reign of terror is installed; thus the landlords and the rich are jailed long before land reforms are put into operation.

The second undertaking is to organize immediately various kinds of unions, such as trade unions and institution unions. The purpose of organizing unions is at least three-fold:

1. The members of a union are organized for indoctrination, with regular study periods for studying and discussing the materials handed down by the Party.

2. Unions are organized as agencies for the administration and controlling of all employment. For instance, as soon as

the union of an institution, whether private or public, is organized, it immediately takes over the administration of that institution. Without the approval of the union no old employee can be discharged nor can any new personnel be employed. Everything is in the hands of the union and the union in the hands of the Party. If a person wants to quit any institution or is discharged, he must get a reference card on which the reason or reasons for his leaving are stated. Without this reference card, no institution, either private or government, dares to employ him. With a reference card having a poor reason for leaving, he is seldom employed. It is so organized that the Party demands absolute obedience from every employee. If the Party decides to discard a person, that person is left to only one way—the way of starvation and death!

3. Unions are utilized as tools by the Communist Party to represent the people in signing manifestoes, in promoting movements, or, in short, in carrying out the orders of the Communist Party. The Farmers Union is organized for the chief purpose of carrying out land reforms, the members acting as witnesses of the people denouncing or condemning the accused at the People's Tribunal.

The third object is to get all the arms and ammunition away from the people. Constantly, for the first week or ten days, the Communist authorities by means of radio, public lectures, posters, newspapers, etc., order the people to surrender their arms and ammunition within a limited time. Beyond that limit, any persons found with even a pistol or hand grenade are, as a rule, treated mercilessly. If there is any slightest suspicion of a person trying to hide some arms or ammunition, he is immediately arrested and locked up.

It is therefore obvious that the Communists leave no chance for the people to revolt.

In carrying out the procedure of the Communist land reform, there are at least six steps, including the actual distribution of land to the poor. They are as follows:

1. INDOCTRINATION OF THE MASSES

This period of indoctrination usually takes from one to three months, depending upon the size of the place and its population, and on how ready the people are to be converted and utilized. Groups of Communist workers and teams of land reform cadres, specially trained, are sent to the place to live among the people, to know them, to help them when necessary in order to win their confidence, and to educate them by personal propagandizing and evangelizing. In addition to personal indoctrination, small study groups are organized for daily indoctrination; union study meetings are held, and mass lectures are given from time to time. During this period, the masses are constantly—several times a day—taught, trained, and mentally "goose-stepped" with the rudiments of New Democracy and Communism, and particularly with the benefits and baits of the land reforms. How glorious will be the days! Work for all, food for all, clothes for all, etc. The poor will be the masters and enjoy the happy days ahead!

The second important task during this period is visitation to the poor farmer by the trained land reform cadres. By "poor farmers" they mean the extreme poverty-stricken tenants and hired farmers, who are called "the property-less." The chief purpose of such visitations is to get from the poor, the downtrodden, and the riffraff their grievances and grudges against the rich, the landowners, the leaders, or any other persons in the community. Wrongs are to be redressed; grievances settled; oppressions revenged, and justice and righteousness to be restored. To the poor, sweet promises are made of a share in the possessions of those found guilty, and of the ownership of a piece of land after the land reforms are completed. It sounds all well and attractive, particularly to the poor and the riffraff. To them the time of comfort and ease is come, and the day of revenge is here! So, grudges and grievances, both true and false, are poured out into the Communist record, including even the grievances of the notorious against the just and benevolent. "The catch is," as my fellow-prisoner and friend said, "that with these records of grievances and grudges, the rich will be killed in order to get their wealth, and the leaders in the community, both good and bad, will be liquidated, because the Communists believe that there is no leadership but leadership of the Communist Party." In the New Regime, doing good is also a sin. This has been confirmed lately by the deaths of a few good and benevolent leaders whom the Communists have come out and called "the benevolent despots."

In addition to mass indoctrination and visitation to the poor, the most important task to be accomplished in this period is the work of organizing the Farmers Union in the district. In their political struggle, the Farmers Union is the instrument and chief means of carrying out their land reforms, including the liquidation of all the enemies of the Communist regime. The Communists are very careful and strict about extending membership in the Farmers Union, not to everyone but only to the poor farmers (the "propertyless"), including the tenant and hired farmers.

After the Farmers Union is organized, the members are indoctrinated with a few more rudiments of New Democracy and Communism, but special stress is laid on training the members in the duties of the so-called masters of the new society, as: how to speak and act at public meetings, how to take the lead in exposing grudges and grievances at the "speak-bitterness" meetings, and how to watch and check others and set examples by bearing witness against them, and denouncing and condemning those brought and tried before the People's Tribunal. They are taught to hate the rich by saying that the rich have enjoyed a life of ease and comfort at the expense of the blood and sweat of the poor. This is the time for them to recover what has been taken away from them, and the day to revenge all the injustice and oppression they have suffered. In order to be the masters of the new society and to enjoy the bright days to come, they must deal with the rich mercilessly and liquidate the enemies of the people themselves. At the first few public trials, they are trained only as choruses, responsive to Communist prosecutors scattered here and there among the crowd, but in the course of time, they soon become the leaders and prosecutors themselves, with others less trained as choruses to give vent to their lowest emotions, by heaping insults on the victims and torturing them, and executing extreme penalties.

2. CLASSIFICATION OF THE PEOPLE

All the people in the district are divided into various classes before land reforms are actually carried out. This work of classification usually takes about three weeks. Based upon the reports compiled by the underground Communist workers, the district registration records, and the investigation of the teams trained for this purpose, classification is first done by a confidential group of committee members consisting of the leaders of the local Party and the loyal and obedient members of the Farmers Union. Later, that classification is presented to a public session of the Farmers Union, specially called for that purpose, for discussion, amendments, and final resolutions. As a rule, what is presented to the Farmers Union is automatically passed with no difficulty, and what is passed at the meeting of the Farmers Union is law.

Roughly speaking, the people in any rural district are divided into four major classes, of which the first two are to be liquidated, and each major class is subdivided into three minor classes. In all, there are at least twelve classes of farmers, or people. They are as follows:

1. *Landlords* are the people who own lands and have never done any manual labor themselves. They live and enjoy life on the rentals collected from the tenant farmers. They are subdivided into big, medium, and small landlords, according to the amount of land they own, the rate of interest they collect on loans to the poor, the total cereal rentals they get from the tenant farmers, and even sometimes according to the proportional severity with which they have oppressed others. In practice, any person who owns twice as much land as the Communists distribute to each poor farmer is generally considered a landlord.

2. *Rich Farmers* are those who own lands and till only a part of the lands they own, and rent out the rest to the tenant farmers. They have to work, but not too hard, to make a living and, in addition, to have some savings to loan to others for interest. They are also subdivided into three more classes, according to the same principles as applied to landlords.

3. *Well-to-do Farmers* are those who make their living by tilling the lands partly owned by themselves and partly rented from landlords or rich farmers, and who also have to supplement their income by doing some other work at the

same time, particularly in the winter months when the harvests are in. Mechanically again, they are subdivided into three minor classes according to the amount of land they own or the total income they earn annually. According to the Communists, this is the major class to which most of the farmers in China belong. The only comparatively tangible principle by which we may distinguish the well-to-do farmer from others is that any person whose income from labor is more than 70% of his total annual income may be classified as a well-to-do farmer.

But he is a landowner, too, so he may be shifted or classified either as a landlord or a rich farmer at the will of the Communist director of land reforms. They are the most unfortunate class, condemned to slow death, while the landlords and rich farmers meet a quick death. According to the latest land law, promulgated on June 30, 1950, the well-to-do farmers are to be "separated" and "protected." The theory of having the well-to-do farmers "protected," even though separated and isolated from others, seems very catching and attractive, but by "separation" they are not permitted to join the Farmers Union. Politically, they are persons deprived of all civil rights. After liquidation of the landlords and rich farmers, "they constitute the only class of people compelled to bear the crushing progressive agricultural tax. Economically, they become, in fact, victims sentenced to slow death."

4. *Poor Farmers* are the real poverty-stricken people who own no land whatsoever and make their bleak living entirely by manual labor. They are the so-called "the property-less" and the chosen people of Communism in land reforms. They include three subdivisions of farmers: (a) the tenant farmers, (b) the annually hired farmers, and (c) the day-to-day hired farmers.

In the strict sense of the interpretation of "the property-less" they are the people without any ownership in this world. The possessive pronoun "my" has no meaning for them. For instance, if a person has two ropes and a bamboo pole, he does not belong to "the property-less" for he still has a little property. Stories are told of persons in the rural districts who are classified not as poor farmers but as rural capitalists because, even though they have no land at all, they have a cow for rent or mule for hire, or three or four quilts to let. The poor farmers are considered the most revolutionary in spirit, for they have no worldly possessions to care for or to hold them back. Consequently, they are utilized in carrying out land reforms, in general, and liquidating landlords and rich farmers and other enemies of Communism, in particular.

Theoretically, there are at least twelve classes of farmers, with the characteristics of each class minutely analyzed, but, in practice, there is no absolute standard or definite criterion by which each class may be distinguished from any other class. Classification of classes is governed only by the principles of Marxism, Leninism, and the New Democracy, just as everything else is governed. In this way the Communists intentionally keep all the privileges and freedom in their own hands in the interpretation of those principles. As a result, the classification and fate of the farmers on the margin between any two of the first major classes depend largely on the impressions of those who constitute the confidential classification committee, the will of its chairman, and the mood of the local party leader in control.

Simultaneously, in this period, another group of the Communist land reform workers undertakes the measurement of all the land in the district. Secretly, they also draw up a plan for allocating lands between those allotted for the experimental collective farms and those to be distributed among the masses of the people.

3. ANTI-DESPOT AND BITTERNESS-TELLING MEETINGS

Usually, two or three weeks are allotted to these meetings in different areas at different times. People are mobilized to attend such meetings; first, by Communist workers, by means of "persuasion" (ceaseless urgings), and, secondly, by orders issued by the Ch'a Chang (leaders of every ten families) and Pao Chang (leaders of every 100 families).

Such prosecution meetings are started with speeches by the Communist leaders assuring the people that every grievance will be redressed, every oppression eradicated, and justice will prevail. To bring about such a reign of justice and peace, every person should and must speak out about his grudges so that he may have redress. After exposing his grievances, every person is guaranteed justice and security by the government. As a rule, the ordinary people are slow to speak but the local Communists, or the trained members of the Farmers Union, "take the lead" or set themselves up as examples by bringing forth planned charges against some landlords, rich farmers, or leaders in the community—charges such as taking rice away from such and such a tenant farmer, forcing the poor to sell their paddy fields at low prices, putting the weak in jail, raping the daughter of the helpless, beating a neighbor's child or dog. Against the wealthy are listed such charges as purchase by force and cheating the poor; against the moralist or the conservative, usually raping or some immoral behavior, and against the leader, often oppression.

To ensure personal safety and security to those who have made public accusations against the landlords, the rich, and the leaders of the community—the enemies of the New Regime—arrest orders are immediately issued and groups of Communist soldiers and local party members, together with policemen, are sent to arrest all the accused who have not been imprisoned by means of the secret-reporting boxes. Thus, another group of people winds up in prison.

Because of lack of space in both old and newly-established prisons, the arrested are divided into two groups, the serious and the light cases. The serious cases remain locked up in the jails, while the light cases are returned to their homes under house arrest, guaranteed by three to five shops or neighbors not to run away. Any person under house arrest must nail a sign above the main entrance of the house—a sign made of a piece of wooden board, three feet square—with the following four characters written in Chinese, "Cruel Despot and Landlord." With such a sign nailed on the front door, the person accused is not allowed to go out nor is any friend or relative of his permitted to enter the house. He is therefore absolutely cut off from all connections in the community. Only with permission from the proper Communist authority is he allowed to go out for some urgent reason, such as seeing a physician. As soon as he gets outside the house he must carry a sign, made of board about three inches wide and one foot long, on which the following is written in Chinese, "I am a cruel despot and landlord." He has to live in shame and disgrace among his friends and neighbors, and no one is permitted to speak to him. For speaking to persons "under control" like him, many people are either put under house arrest or jailed immediately.

A number of neighbors are assigned to watch him and check constantly. Any suspicious actions of his must be reported immediately, in addition to his own weekly report about himself and that of the leader of the controlling neighbors. The door of his house has to remain open at night as well as in the daytime. The apparent reason for this is so that he cannot do anything reactionary but, in fact, it is to give the Communist workers the opportunity to make the so-called "visitations" at night. Visitations at night are ceaseless inquiries and trials, in order to exhaust him by giving him neither rest nor peace.

4. THE PEOPLE'S TRIBUNAL

Disgorging of rents, confiscation of properties, and public trials are the three phases of the People's Tribunal, which is considered the final battle waged by the poor and oppressed against the landlord. This takes a couple of weeks. By this time the poor have "progressed" enough to see how they have been oppressed, directly or indirectly, by the rich, and their hatred against other classes has been instigated and well developed by the Communist indoctrination. So it is no longer difficult for the Communists to persuade the poor, especially the riff-raff, to do what they want.

1. *Reduction of Rents and Disgorging of Rents*—It is nationally recognized that the former rentals and usuries were too high and oppressive to the poor. In former days, the tenant farmers had to pay their landlords an average of fifty to sixty per cent of the main crops, and usuries in the rural districts were as high as from three to ten per cent monthly.

When the policies of reduction of rents and return of loan securities were announced, they were warmly welcomed, not only by the poor but also by many liberal-minded land-owners and bourgeois, but no one had ever dreamed of the necessity for reimbursement of the overcharges in interest and rentals for so many years past. Furthermore, there is no standard for the reimbursement of rentals and usuries. Some are ordered to reimburse 25%, others 50%, still others 100%, 150%, 170%, and even several hundred per cent.

The regular procedure is that the percentage of reimbursements is first decided by the Communist workers together with the local members of the Farmers Union and, later, this is presented to the mass meeting, with the obedient members scattering themselves among the masses to help the pre-arranged percentage get passed. Thus, they say it is passed by the people and at the people's request!

Again, the number of years for such reimbursements is likewise decided; some ten years, others fifteen years, still others twenty years, and sometimes, in not a few cases, even up to fifty years past, with compound interest. A great "catch" in this step is that no reimbursement is made directly to the poor, but is made to the government. Although there is no standard, there *is* a fixed principle by which the Communists decide the percentage and the number of years of reimbursements of rentals and interest. They decide so much percentage and so many years that the landlords or the rich, in every case, are not able to pay even if all their properties, both movable and immovable, were sold at their proper value! No one has been able to make such reimbursements. Consequently, comes the second phase of the work, called:

2. *Confiscation of Land and Properties*—Land is taken by the government for the future collective farms and for distribution, while properties, so confiscated, including houses, money in gold, silver, and bank notes, clothes, furniture, agricultural implements, domestic animals, etc., are disposed of as follows:

Furniture, clothes, and domestic animals are divided sometimes as free gifts or rewards to the poor who have been loyal and obedient. At other times, when the Communists want to keep them for their own use, they set "low" prices on them for the people to buy them back. No matter how low the prices are set, the poor farmers, as a rule, are unable to buy. Then, without exception, some loyal member of the Farmers Union will come forth and move that they should be stored at the government office for the time being.

All the agricultural implements, cows, and water buffaloes become, without question, the property of the Farmers Coöperatives (supposedly of the farmers, but actually of the government), to be rented to the poor peasants for use in tilling the land. The charge for the use of such implements and animals is thirty per cent of the harvest, which is one form of the Farmers Loan.

All the money in gold, silver, and U. S. banknotes, forbidden to be circulated, is naturally turned over to the government.

3. *Trials by the People*—These, generally speaking, may be divided into two classes, the smaller designed for the purpose of humiliation and submission, and the larger for liquidation. The former is not so well organized as the latter. We may call the smaller the Public Trial and the larger the People's Tribunal. In Chinese, both are called by the same name, "Kung Sheng." The public trial is carried on mainly to humiliate the intellectual, and leads often to house arrest or imprisonment, while the purpose of the People's Tribunal is two-fold: to liquidate the accused and to threaten the people attending the trial. As we Chinese would say, "To

kill the chicken is to threaten the monkey." During the time of land reforms, the object of the People's Tribunal is to liquidate the cruel despot and landlord (imperialist). Those two names always go together in accusing any landlord.

The People's Tribunal is often held in a big public ground, where a temporary stage is built, or in a temple courtyard where, usually, a stage has been attached. Across and above the stage is a long piece of red cloth with the following inscription written in yellow Chinese characters, "The People's Tribunal of such and such person, a cruel despot and landlord, of such and such village, such and such county." Papers of various colors are all around the stage, inscribed with such catching and emotion-arousing slogans as, "The final settlement with the cruel despot and landlord"; "The landlord is the sucker of the people's blood"; "To kill the landlord is to emancipate the peasant!"

About a week before the trial, the Communist Party gives orders to the Farmers Union in the village where the trial is going to be held and also to the unions in the surrounding villages. After having received these orders, the authorities of the unions, usually about three days before the trial, give orders to every person in the allocated area to attend the tribunal at such a time, on a certain day, in a certain place, without fail. Any delay or absence is checked and punished by the Farmers Union. In groups led by the leaders of the Farmers Union, the peasants of the various villages arrive at the People's Tribunal punctually. To each group a number of the local Party members are assigned, whose duty is to tell every person in the group to respond, as the member of a cheer team, and repeat what the leader says or shouts, and then to check those who try to be indifferent or keep their mouths shut. In addition, there is a general leader, a very well-experienced Party member, standing in front on one side, who serves as a general prosecutor and director, from whom the leaders of the various groups get their directions for action.

As various groups are marching into the Tribunal ground, the judges—made up of the leader of the village Farmers Union, the village elder, representatives of farmers and laborers, with the chairman of the particular village Farmers Union as the chief judge—take their seats behind a long table near the front of the stage, which is also crowded with many Communist assistants. The accused, as a rule handcuffed and his body tied with ropes, is brought before the judges on the stage by guards. Soon the People's Tribunal starts with a short, coached speech by the chief judge, denouncing cruel despots and landlords as the enemies of the people. After the speech, he asks the crowd below whether or not they want this particular cruel despot and landlord to make a confession himself.

Quickly the leaders of the group respond, "We want it," which is followed, also quickly, in a thunderous voice from the crowd, "We want it!" Thereupon, the accused is ordered by another judge to make an utterly frank confession of his past sins.

Having been imprisoned for months, and suffering from hunger, thirst, dirtiness, endless interrogations, sleeplessness, restlessness, nervousness, threats of torture or actual tortures, he has been constantly indoctrinated and taught that grace can be granted only to those who want to redeem themselves by means of frank confessions. Hoping to obtain the promised grace, the accused makes a confession in a general way, not caring whether it is true or false, by saying, "I am a landlord. I eat, but don't do a thing. My comfortable life is built upon the exploitation and oppression of the tenant farmers. I treat them like slaves. Once in a while I scold them and sometimes I beat them up. I deserve the punishment I am getting."

After his confession, the chief judge again asks the crowd, "Is he frank enough in his confession?"

The leaders answer, "No, he is not frank enough."

"What should we do with him, then?" asks another judge.

The leaders shout, and the crowd repeats as a chorus, "Beat him, beat him." Immediately, the Communist assistants on the stage take out long bamboo sticks or clubs which have been prepared beforehand. Then he is beaten mercilessly, with blood flowing down all over his face and hands. This time, the general director in the crowd takes over by shouting, "Make him kneel down on the broken glass; we, the people, want to try him." Some local militia come forth with a bag of broken glass and porcelain and pour them on the floor before the judgment table. He is then forced by the guards to kneel down on the broken glass facing the people. In a few seconds, his legs and knees are cut up and more blood drips down.

"Now," announces the chief judge, "any persons who have grievances against this cruel despot and landlord may come up on the stage and declare his sins before the people. Grievances will be revenged, blood for blood!"

Possibly one or two of his enemies are pushed forward to make accusations, and they accuse him of some petty misdoings such as scolding Shen Ta Shao, wife of one of his tenant farmers, or demanding so much rice as rental for land after a flood. For fear the emotion of the masses will cool down, a well-trained Party member comes forth whom the accused had never met before. Coming up on the stage, the coached prosecutor slaps first the right side of his face and then the left, as hard as he can, and accuses him of a long series of crimes, rolling one after another out of his mouth in an angry tone, such as, "He does nothing except eat the most delicious dishes, drink famous wines, Mao Tai and Fang Chiu, and smoke opium every day. He demands high interest on loans to others. He compelled so and so to commit suicide and raped such and such a woman. He has always oppressed the peasants and is a great enemy of the people. This cruel despot and landlord should be killed by the people!"

At this point the group leaders yell out, "Kill him! Kill him! Shoot him!" So repeat immediately the masses, "Kill him! Shoot him!"

Then the chief judge gets up and asks the crowd, "Do you really want to kill him?"

"Yes, we really want to kill him," respond the leaders and the crowd, as if thunder were roaring all over the place.

A few seconds later the sentence of death, well prepared and written beforehand, and sealed with the red seal of the People's Court, is pulled out and read by the judge, declaring that the government has accepted the opinion of the people and that this cruel despot and landlord, so and so, is to be shot before the people. After hearing this sentence the accused, terrified beyond his expectations and dreams, faints and falls down on the stage and his facial color changes from pale to green. In his semi-consciousness, he is dragged down from the stage by the local militia to an open space behind the people. He is told to stand but he cannot do it. While he is half kneeling and half falling on the ground, he is shot three times from behind. So ends the life of one of the 14,000,000 landlords of China who are to be liquidated.

According to the Communists, the landlords in China—about three per cent of the total Chinese population—are only one class of enemies of Communism. "The People's Tribunal," according to one of the leading Communists, "is the final bitter battle against the enemy of the people" ("people" meaning Communism). Any person brought before the People's Tribunal is surely doomed. All sorts of accusations, some true and most false, are presented by the trained prosecutors and, if possible, by the enemies, servants, or relatives of the accused. Certain people, the riff-raff of the community, are coached beforehand and planted among the masses to go up on the stage to spit on the accused or to beat him with their fists, bamboo sticks, or clubs. Tortures such as stripping him to his underwear and dipping him into a pond in winter, or making him kneel on broken glass and porcelain, are often applied before he is condemned to death either before a firing squad or by being stoned to death by the populace. Death is the end because the Communists believe that death, on one hand, gives no chance for revenge and, on the other, accomplishes the first step toward their classless Utopia.

5. DISTRIBUTION OF LANDS

After liquidation of the landlords and the rich farmers, and isolation of the well-to-do farmers, practically all the lands are in the hands of the government. At a big ceremony, the property deeds are burned up; the people are jubilant, and there are songs, dances, and parades in the expectation of receiving the promised land! The lands so acquired by confiscation are divided into two portions, one for the government experimental collective farms and the other for distribution. The land for distribution is again divided by the Communist authorities into three grades; the first grade to be distributed among the members of the Party, particularly those who participated in the 25,000 *li* (*li* is about one-third of a mile) march from Kiangsi to Yenan, Shensi in 1934, the second grade for the family members of the soldiers fighting at the war front, and only the third and worst to be distributed among the poor.

Before the Communists took over the mainland, they had promised to give each person at least three *mow* of land (one-half acre) but, at the time of distribution, the amount of land given to each person varied at different places, although the general principle is that the minimum of one *mow* (one-sixth of an acre) to the maximum of two *mow* (one-third of an acre) has been distributed. With that small piece of land, the poor peasant began to realize the small chance and, in fact, the practical impossibility of making a living, but it was too late and he had to make the best of it by working harder.

Another factor which created an enormous amount of fear and scepticism in the minds of the farmers was that in getting lands from the government, many families were broken up and scattered. They heard, and saw, that while the members of many families received one to two *mow* of land each, they were not in one locality, but scattered in different counties or districts. Thus, the members of the same family could not live together or work together, and the family was broken up. Although they had been good, even while being poor, this might be a sort of punishment for those who had kept quiet about denouncing the landlord or had shown no zeal for the new regime! The family tie is strong in China! So all the rest feared and worried. Perhaps some day, in the same way, their families might be broken up, too, as they were given no deeds to the distributed lands!

Except for the well-to-do farmers, who constituted a separate and ostracized class by themselves, all the other peasants who have received a piece of land from the Communists are poor and have never owned any farming implements. They were all supplied before by the landlords or the rich farmers, but now all the farming implements, cows, and water buffaloes of the landlords and the rich have been turned over as properties of the Farmers' Coöperatives. The peasants cannot work on the land with their hands only, so, in order to till the land, they have to go to the Farmers' Coöperatives and sign an agreement that for the use of those implements, thirty per cent of the harvest is to be given to the Coöperative. This is called the Farmers Loan. In spite of all these difficulties—the worst he has ever faced—the farmer, still hoping for the best, gets up earlier and goes to bed later, and works harder, day in and day out, with the hope that agricultural production may be increased so he can feed his family, and that the new government, after the storm, may realize the sufferings of the peasant and improve his condition.

At harvest time, the Communist rice collectors arrive in the district. Mass meetings are called and personal visitations made. The farmers are told that they should be happy and grateful to the government for the ownership of the land given to them and for liquidation of the landlords, to whom they need no more give fifty to sixty per cent of their crops (formerly, the tenant farmer was required to give from 50 to 60% of the main crop—rice in the East, South and Central China, and wheat in North China). Now the government is so lenient with them that they are required to give to the government only twenty percent (during the first two years it was only thirteen per cent but has been raised to twenty per cent)! They are also told that they are now masters, not slaves as before, and they should act as masters, and feel privileged to present twenty per cent to the government, happily and gratefully.

The obedient farmers are, of course, happy to express their willingness and gratitude by nodding their heads, but soon their hearts are aching when they hear that the twenty per cent is 20 per cent of *whatever* they raise, including all kinds of cereals, cattle, sheep, chickens, swine, etc., and even their fuel. Their hearts are totally broken when they find out the second "catch"—that the twenty per cent is not of the actual production but of the estimate of each made by the Communist collectors in the area. For instance, one *mow* of land in China produces between 250 and 300 catties (one catty is slightly heavier than one pound). Before 1952, the official estimate was 323 catties but since then, it has been 570 catties. Again, the unofficial estimate made by the local Communist collectors varies in different places, but is always more than the official estimate. Only by doing this can the local collectors fulfill the requirements and, furthermore, earn more merits for themselves. It is an expression of efficiency and loyalty at the expense of the poor farmers.

Naturally, the peasants were downhearted and discussed their unfortunate fate, with deep sighs, and some brave ones among them, individually or in groups, even protested to the collectors against such unreal and unreasonable estimates. Such protests were always turned down, on the infallible ground that they were no more slaves, but masters of their own lands; they should work harder than before and must increase agricultural production. Only the lazy cannot bring the production to what it ought to be!

Lately, in some places, in order to simplify the matter, instead of making estimates of this and that cereal, of chickens and eggs, and what-not, the Communists have instituted a single Agricultural Tax of forty per cent of the estimate of the main crop.

After having paid all the required taxes—twenty per cent of the estimate of whatever they raise or forty per cent of the estimate of the main crop according to the Agricultural Tax, and thirty per cent of the estimate towards the Farmers Loan for use of the farming implements, etc., in addition to the forced contributions for the war front or for some relief purposes, and the compulsory purchase of victory bonds— the farmers, although having worked much harder than ever before, are left, if not totally rice-less, to face hunger or starvation for at least a large part of the year.

Since the institution of land reforms, another law which has been enforced very strictly throughout the country is that every farmer is allowed officially 500 catties of rice a year for his own living, including his clothing, medicine, and what-not. However, the prevalent allowance in the rural districts is 300 catties of rice per person per year. Any surplus rice must be sold to the government at the government-fixed price. The law also requires that even the 300 catties of rice that each farmer is entitled to have cannot be kept at home but must be stored at the government granaries, guarded all the time by the Communist soldiers or the local militia. The local government issues a certificate of ownership to each for the 300 catties of rice with which he may go to the specified granary every three or five days to get his supplies.

The real purpose behind this is that if, in the meantime, any farmer has been reported for having said or done something or even grumbled against the new regime, he will no longer get his own rice from the government granary.

A true story of real ingenuity may be told here. A clever farmer named Chang Shan (excuse me for this false name for he is still behind the Bamboo Curtain) lives in a certain village in Central China. He has a lovely family, including an understanding wife and two devoted children, a son of nineteen and a daughter of seventeen. They worked hard together, with neither a tenant nor a hired farmer to assist. Every year they were able to save a little, with which they bought a few *mow* of land. At the time of the land reforms he was considered a well-to-do farmer and so has been ostracized—unable to join the Farmers Union yet bearing the crushing burden of the progressive Agricultural taxes.

At the completion of land reforms in the district, Chang Shan knew what would happen to the rice he had stored at home for rainy days. Most probably it would be confiscated and he would be punished—he had one relative in the neighboring district who had paid the extreme penalty of five lives (all the members of that family) for storing only five *tu* (each *tu* has about fourteen catties) of rice at home without reporting to the government. He himself had about ten times that much rice at home and was therefore ex- tremely afraid, but unwilling to give up the rice.

He thought and thought; for nights he could not sleep. At last he hit on an idea and consulted with his wife and chil- dren. For two nights in succession he blacked out the win- dows and closed the door. The four of them pulled the big stone family grinder by turns and had all the reserved rice— five *tan* in all (each *tan* has 140 catties)—ground up into rice powder. On the third night they worked together again; one carried water, the second mixed up the rice powder and water into a paste, the third and fourth, as bricklayers, put the rice paste on both sides of the bamboo partitions inside the house like plaster.

One week later, true and sure, a group of five Communist inspectors, armed with pistols, came to his house. They searched and searched everywhere inside the house and even dug up about one-third of the floor in Mr. and Mrs. Chang's bedroom (the only floored room in the whole house) but they could not find a single grain of rice! They were bewildered and angry with him. However, at last they left! Since then, whenever they cannot get rice or other foodstuffs to eat, the family just pull down a piece of the rice "plaster" at night, steam it in the dark and help themselves!

6. LAND REFORM INVESTIGATION

The objective of land reforms is "not so much to divide and give the land to the poor as to liquidate the landlords and all the opposing forces" to the Communist regime, including even tenant and hired farmers. The editorial published in the *Communist Southern Daily* of March 20, 1952, said, "Many poor tenant and hired farmers, bought with money by the landlords, have continuously and purposely opposed land reforms and planned anti-revolutionary plots and movements. . . . We, the Government of the people, must also thoroughly eradicate all those tenant and hired farmers who are willing to be oppressed and utilized by the cruel despots and landlords." The objective of land reform investigation is just the same—"to kill and kill more, not only those who are actively opposing the Communist regime but also those passively resistant and non-coöperative." It makes no difference whether rich or poor.

According to the Communist directives, the objective of

land reform investigation is threefold: (1) to investigate the "black" or hidden lands; (2) to investigate the landlords who have been overlooked, and (3) to investigate the mistakes and faults of the former land reform workers. What land reform investigation gives to the farmer is more "struggles," more sufferings, and more deaths. During the period of land reform investigation, many farmers are promoted or, as the Communists say, readjusted or corrected, in the classification of classes. Many poor peasants, formerly classified as tenant or hired farmers, are now readjusted as well-to-do farmers. Many formerly put in the category of well-to-do farmers are now promoted to be rich farmers or landlords. Thus, the landlords are changed into cruel despots, to be immediately liquidated. In other words, the higher you go in the classification and reclassification, the nearer you approach to your death!

Not only are classes of farmers reclassified but the agricultural taxes also have to be corrected or readjusted accordingly. This readjustment is forcing thousands of farmers in the rural districts to commit suicide, in prisons as well as at home, and compelling millions of them to forsake their dear good earth homes and flock to the cities, as is being done to-day, to try other means of earning a living. We all know well that it is not easy to get the farmers to forsake the good earth. When this happens, they are suffering more than they can bear. It is a sign of rural bankruptcy—and will mean the future bankruptcy of the Communist regime in China.

After the completion of land reforms, the land reform investigation teams are constantly sent all over the country by the Communist government in Peiping to receive more secret reports, to examine the classification of classes of people, particularly that of the well-to-do farmers, to check the efficiency, orthodoxy, and thoroughness of the former workers, to hold more People's Tribunals, and to liquidate more enemies of Communism. For fear that the sympathy of the old land reform workers might be aroused for the poor masses of the people so that they might coöperate with them in possible future resistance or opposition to the Communist regime, the land reform investigation teams, consisting only of newly-trained youngsters (who know nothing of the sufferings of the poor but only about orders "from above"), are sent to take their places for the work of more investigation, more land reforms, more "struggles," more confiscation, and more liquidation. They will be followed by other investigations called Fu Tsa (re-investigation), Shan Tsa (third investigation), and so on, one after another, until all the wealth and lands are in the hands of the State, all the opposing forces against the Communist regime liquidated, and all the poor absolutely dependent upon, and obedient to, the Government! It is not land reform, but land reforms without end!

To illustrate the work of the investigation teams, another story may be told, the story of an old teacher of mine whose name is Ching (excuse me again for this false name; it is for the protection of his family). He was a typical Chinese scholar of the old school. Teaching the ancient classics was his profession. Before the establishment of the Republic he had his own tutoring school and, after that, had taught at the Government and private schools. For thirty-nine years he did nothing except teach the Chinese classics. His students are found all over the country, and all remember his lovable character, his patience with pupils, his quiet conversation and slow movements. His only desire in life was to become the teacher of teachers and, to a great extent, he attained that goal.

Little by little, he accumulated some savings; year by year, he acquired some land. At the time of land reforms in 1951 he owned a total of 70 *mow* of land (about 12 acres). As a scholar and teacher, he had heard and read a lot about the Communist reforms and while he knew he had no chance of keeping his land, he was not afraid of the land reforms. Often he consoled his friends and himself by saying, "I have never robbed any person or done any harm to anyone, or taken part in politics or revolution. All the land I have has been bought with my own savings from teaching. Why fear? At most, they may take the land but not my old life!"

At the height of the Communist propaganda for land reform in his district, he found that the director of land reforms was one of his former naughty students. It happened that, somehow, one night, Mr. Ching was able to see the director personally at the latter's home. During the conversation, the student expressed his concern for his old teacher and agreed finally to take over his 70 *mow* of land peacefully, without "struggle" or liquidation, and even give him a share of two *mow* of land. Although it was pretty hard for Mr. Ching to give up his own property, after having saved for thirty-nine years to attain it, he realized from the conversation that night that his student was doing him a great favor by accepting the land "peacefully." Occasionally he comforted himself by saying, "Well, this is the time and tide; I am satisfied, and expect to continue my teaching 'til I die. Now I am a son of Mohammed—I was born naked and will die naked!"

Soon the land reforms were completed in the district and a report was submitted by the director to his superiors. In the report, Mr. Ching's name and land were mentioned vaguely, but nothing about the "struggle" or liquidation of Mr. Ching, the landlord. Then, many weeks later, a land reform investigation team arrived and the members of the team paid special attention to Mr. Ching's case. It was easily found out that the director had taken Mr. Ching's land by the unorthodox method of no "struggles" and that Mr. Ching, a typical landlord, was even given a share of two *mow* of land and was enjoying life. Consequently, the director was summoned to the group criticism meeting and severely rebuked by the investigation team. He was told, "You are not worthy to be a member of the Party. The Party trusted you and gave you this important work, but you were not faithful. This was your chance to express your gratitude to the Party and to earn for yourself a big merit. Instead, you have committed two great sins—the sin of coöperating with a reactionary and the sin of sentimentalism. You don't know how to carry out land reforms! You had better go back and study some more!!" Next day, he was degraded and sent back, under guard, to the so-called Revolutionary College, where he was required to purge himself of his sins by studies, self-criticism, and hard labor.

In the meantime, the local members of the Party and the Farmers Union were called together, and Mr. Ching was arrested and imprisoned. Five days later, a People's Tribunal was held and old Mr. Ching, tied and handcuffed, was brought forth and tried by the orthodox method of "struggles." After four hours of humiliation and torture, he was condemned to pay the extreme penalty. Before the people, and by the people, he was stoned to death!

The Real Objectives of the Communist Land Reforms

Mao Tse-tung, in one of his printed speeches, said, "Communism in China has three difficult 'passes' which the Communists have to fight through: (1) to win the civil war and

take over the government (which they finally did in 1949);
(2) to carry out land reforms successfully, and (3) to change
the temporary State of New Democracy to the Communist
State," when the State is supreme and owns everything,
even including the very life of each person. The most important of the three is the land reforms, which is the key to
understanding Chinese Communism and success in getting
over the first 'pass'; also, to success, as they hope, in getting
to the third eventually.

Of land reforms, Mao Tse-Tung and his Communist colleagues have repeatedly said that the purpose "is to emancipate the enslaved, to increase agricultural production, and to
improve the living conditions of the poor." It sounds very
sweet and attractive! "And they know very well that there
would be more production if they were willing to utilize
their organizational discipline and efficiency to follow the
path of peaceful reformation instead of using the method of
brutal 'struggles' and liquidation. Knowing this and seeing
that the result is often not increase but decrease of production, they still insist on the 'orthodox' method of carrying
out land reforms." It is evident that there must be some other
important objectives under the much-pronounced sugar-
coated aims.

From the confession of a leading Communist (Yeh Chi
Sui's article published in the *Communist People's Daily*,
Peiping, July 18, 1950), which openly declares, "The central
object of our work in land reforms in the past has been to
win the revolutionary war—to mobilize the immense human
and material resources of the village," we can get some light
and readily see that the chief objective of land reforms prior
to the establishment of the Communist Regime was to
mobilize the human and material resources of the villages

in order to win the civil war and take over the government. This has been verified.

As I have said before, in the distribution of land each peasant was given one to two *mow* of land, but with that small piece of land, it was practically impossible for any farmer to exist under any standard of living without going hungry or being compelled to eat wild vegetables, fruits, grass, etc., especially during the season just before the annual harvest, to say nothing about the Agricultural Tax and the Farmers Loan. By this means, plus the heavy taxes, forced contributions, and unlimited confiscation, the independent living of the farmers was wiped out completely. At the same time, they saw the Communist soldiers, well fed and clothed, stationed in their villages, and it was only natural that they would throw away their plows and join the armed forces. Furthermore, in the process of liquidation of the landlords, the rich, and many of their fellow-peasants, the poor farmers were utilized as instruments of humiliation, torture, and death. So, "in the depths of their hearts they were troubled, their consciences bothered them, and their fear of retaliation and revenge made them constantly seek some refuge. After much thinking, they all came logically to the conclusion that, in order to survive at all and cover up their past crimes and sins, the only way for them was to join the army."

In short, the chief objective of land reforms between the years 1927 to 1947 was to force every peasant to join the Communist forces so as to win the civil war and take over the government. It worked like magic; the peasants just swelled the Communist army as volunteers so that it grew by leaps and bounds. So the "human sea" strategy was adopted, and they took over the mainland in 1949.

Even today the chief objective of land reforms, followed by checking of the investigation teams, remains the same, i.e., "to mobilize the immense human and material resources" of the whole nation for the larger purpose of winning the revolutionary war in the world. By means of secret reporting boxes, arrests, liquidations (including the slow death at slave camps), the so-called reform by labor, disappearances, murders, and the People's Tribunal, all the enemies of Communism, including the landlords, the rich, the leaders, the educated, the disobedient, the indifferent, etc., are quickly destroyed. Thus, to consolidate their power of control and to accomplish the first step toward the Communist State, at least, if not the ultimate Classless Utopia, this is the third "pass" they have to fight through.

In the same way today as they did a decade ago, the Communists in China, by means of merciless confiscation, forced contributions, multitudinous crushing taxes, distributing inadequate land to the poor, and controlling all the foodstuffs, are attempting to enforce their "hunger policy" by which the State is to be enriched and strengthened, while the people are made poor, dependent, and obedient. This leaves the people only one path, the path of joining the Communist forces, either as soldiers fighting the war of world revolution or as watchdogs or policemen doing the work of watching, checking and reporting, if they want to survive at all. Otherwise, they will be prisoners, locked up in jails or working in slave camps, to be sooner or later liquidated. Thus the Communist conception of "State", defined as consisting of only three groups of people—soldiers, policemen, and prisoners—is fulfilled.

As a conclusion, the words spoken by the second leading Communist in China, Liu Shao Chi, well express the real objectives of land reform: "Land reform is an organized bitter struggle. Its chief aim is not so much to divide and give land to the poor as to destroy the feudalistic forces," i.e., the enemies of Communism.

From this analysis of the procedure of the Communist land reforms and the statements and confessions of the leading Communists in China, we can readily see that the Chinese Communists never have been and never will be agrarian reformers, but are part and parcel of world Communism for world revolution, as clearly and frankly stated in almost all Communist books in Chinese.

The Communist Version of Eminent Domain

It has been learned from a reliable source that the Communist regime in China issued an edict on December 15, 1953 that, hereafter, the Central Communist Government may, without compensation, confiscate any amount of land for the purpose of national reconstruction and remove the inhabitants thereof, in any number and at any time deemed necessary. This reconstruction covers factories, mines, railways, communications, water works, municipal reconstruction, and all other economic and cultural construction for national defense.

When it is found necessary to use more than 5,000 *mow* (one *mow* equals 1/6 of an acre) of land or remove more than 300 families of the inhabitants for local reconstruction purposes, the Area Government, such as East China, Central and South China, and Southwest China, may issue the order. In the case of less than 5,000 or more than 1,000 *mow* of land, and less than 300 or more than 50 families, the Provincial or Municipal Government has similar authority, while for less than 1,000 *mow* of land or 50 families, the District Government (similar to county in the United States) is authorized to act.

It is further proclaimed that, in the time of emergency, any amount of land may be confiscated and any number of families removed without reporting to, or obtaining approval from the proper government beforehand. The order may be given and filed at the time of confiscation, or afterwards.

Another point of significance in the law is that any land so confiscated for national reconstruction or defense, if not used, shall be held by the government and not returned to the individual owner or sold to others.

There is, however, one "sugar-coated pill" in the law. Except for the vacant lands and those of the landlords so confiscated without any compensation, some subsidies may be given for the rest of the condemned land to pay for what is built or raised on it. A meeting is to be held to decide the amount of subsidies. Those attending consist of the authorities of the local government, members of the organization for which the land is to be condemned, representatives of the Farmers Union, and the owner or user of the land. This looks rather democratic, to have the matter decided by the four groups concerned, but when the vote is taken as to the amount of subsidies to be given, it is always three against one—with the original owner or user of the land in the minority. Thus, the promised subsidies will amount to nothing. This law, therefore, means not only the confiscation of land but also includes whatever is built or raised on the land.

From a close study of this version of Communist Eminent Domain, we can easily see that the purpose of this law is at least twofold—"killing two birds with one stone." On the one hand, the Communist regime is preparing energetically to meet the forthcoming invasion of the mainland of China. It authorizes the local authorities, both civil and military, to build the necessary defenses and to remove the people so that they cannot co-operate with or help the invaders.

On the other hand, the so-called transitional regime, so often named by the Communists as the "Reign of New Democracy," which has permitted the national capitalist and the bourgeois to exist and which distributed a very small piece of land (from one to two *mow*) to each person as a token ownership—in order to lull the people—is fast disappearing. As we Chinese would say, "The fox-tail of the Communists behind the Bamboo Curtain is now clearly exposed."

The Communists in China no longer care about fooling the people and hiding their real intentions and objectives in regard to land reforms. Drunk with power, the Communist regime in China has killed 43,400,000 and controls more than 7,000,000 innocent Chinese as slaves. They think they have consolidated their power; now, they will abandon the "State of New Democracy" and, by force, usher in the Communist State.

There will be no more private ownership of land in the near future but collective farms, with all farmers as their slaves. More millions will be killed; more millions will flock into cities for other means of survival; more millions will flee into the high mountains to join the guerilla forces! More unrest and more revolts! Consequently, although they do not realize it, the Communists are digging their own graves!

This is the greatest calamity that has happened to the Chinese race, and should be a very good lesson to the peace and freedom-loving people elsewhere. May God bless the enslaved with courage and give them opportunities to regain their freedom! And may He bless us who are free, with wisdom and understanding not to be fooled any more by the dialectical and diabolical evils of Communism!